RICHARD OVERY

THE PENGUIN
HISTORICAL ATLAS
OF THE THIRD REICH

PENGUIN BOOKS

Published by the Penguin Group
Penguin Books Ltd, 27 Wrights Lane, London W8 5TZ, England
Penguin Putnam Inc., 375 Hudson Street, New York, NY 10014, USA
Penguin Books Australia Ltd, Ringwood, Victoria, Australia
Penguin Books Canada Ltd, 10 Alcorn Avenue, Toronto, Ontario, Canada M4V 3B2
Penguin Books (NZ) Ltd, Private Bag 102902, NSMC, Auckland, New Zealand

Penguin Books Ltd, Registered Offices: Harmondsworth, Middlesex, England

First published 1996
3 5 7 9 10 8 6 4 2

Text copyright © Richard Overy, 1996
Design and maps copyright © Swanston Publishing Limited, 1996
All rights reserved

The moral right of the author has been asserted

Printed and bound in Great Britain by The Bath Press, Avon

ISBN 0–14–0–51330–2

Foreword

In 1938 the German government monthly statistical journal *Wirtschaft und Statistik* published two maps on the Jewish question. The first showed the spread of Jewish populations in Europe, with a dark-shaded mass covering the whole area east of Germany as far as Moscow and the Crimea, hanging like a menacing shadow over the new Reich. In the Nazi period mapping mattered a great deal. Germans mapped races and racial movements. They mapped territories. There could hardly have been a schoolboy or girl in Germany who could not recite the territories taken from Germany after 1919; the Caroline Islands, the Marshalls, Samoa, Togo, Cameroon and so on. When Germany was master of most of Europe Nazi planners used mapping as the base of their vision of a German-centred Europe. One of the most famous propaganda maps of the period shows Nordic Europe threatened on every front by inferior races — semites, Slavs, blacks and, more surprisingly, Latins. Under such circumstances it is more than appropriate to view the Third Reich through the lens of an atlas, and some of the maps that follow have drawn on the German inter-war obsession with cartography.

Even in an atlas devoted to only a dozen years of history, a remarkably short period, given the huge weight of historical significance the Third Reich is made to bear, it has not proved possible to include everything that could be mapped. The atlas is not intended as a geographical guide. Population density or the locations of German mineral ore deposits can be found in economic geography texts of which N. Pounds, *The Economic Pattern of Modern Germany* is an excellent example. The maps and charts have been chosen because they tell the story of this dozen years effectively in atlas form. Issues of race, area and resources were, as we have seen, central to the Nazi view of Germany's future. Even after the war the territorial dimension lived on in the desire for German re-unification, or the demands of the radical right for the recovery of the 'frontiers of 1939' or, very occasionally, the frontiers of the older Reich of 1914.

In constructing the atlas wide use has been made of published material. References at the end of the atlas indicate more precisely that debt, but a general and grateful acknowledgement is not out of place here. I also owe a debt of gratitude to Andrew and Ailsa Heritage for their editorial work on the atlas, and to Malcolm Swanston and Andrea Fairbrass who have had to turn my rather imprecise conception into finished artwork. Any errors that remain are mine. Finally, thanks as always to my growing family.

Richard Overy
King's College, London, 1996

Contents

I: From War to Third Reich 1918–1933

In 1914 Germany was a wealthy empire. The First World War brought defeat, impoverishment and a parliamentary Republic.

"Life was madness, nightmare, desperation, chaos…"
Erna von Pustau
(on inflation)

On 18 January 1871 in the Hall of Mirrors at Versailles the king of Prussia, Wilhelm I, proclaimed the establishment of a German Empire. The new state drew together in a federal structure all the smaller states of the German Confederation and Prussia. Austria, which had dominated the much looser Confederation from its foundation in 1815 until her defeat by Prussia in 1866, was excluded. The German Empire constituted what was known as 'lesser Germany' (*Kleindeutschland*). Greater Germany, a German empire which included Austria, had to wait until Hitler, Austrian-born, brought the two together in 1938.

The new German state was dominated by Prussia and its Minister-President, Prince Otto von Bismarck. He became the Chancellor of the Empire, and the moving force of its politics. He succeeded in creating a nominally parliamentary state in which the final say remained with the Chancellor and the Emperor. For the Empire's entire life down to 1918, when it was destroyed by defeat in war, the traditional ruling-classes succeeded with growing difficulty in maintaining a system in which they enjoyed a disproportionately large voice and sufficient political strength to prevent a full-scale democracy. The structural imbalance in favour of the old elites, sustained by institutions such as the imperial army, was challenged by popular political forces from the 1870s onwards. Liberals and social-democrats campaigned for a genuine democracy. The Catholic minority in Germany became organized in one of the most successful mass parties, the Centre (*Zentrum*). By 1914 a new right-wing mass had emerged which was also impatient with the survival of the old Prussian-centred order, and preached a blend of popular nationalism, racism and anti-Marxism. It was among the political circles of the radical right that the roots of Hitler's National Socialism were to be found.

Berliners scavenge on a rubbish heap in Berlin in 1919. Shortages of food and goods created great hardships at the end of the war, but for many Germans poverty was a fact of life throughout the inter-war years.

Before 1914 the imperial system set up by Bismarck had already changed a good deal. In 1870 two-thirds of Germany lived and worked on the land. By 1914 the figure was only one-third. Industrial modernization made remarkably rapid strides. The German economy grew at almost 3% a year from 1871 to 1914, and its export trade grew even faster. By 1914 Germany was the world's second largest industrial power behind the USA, and was closing rapidly on Britain as the world's largest exporter of manufactured products. The growth of modern industry and technology created social forces which could not easily be contained within the traditional political power structure. There existed an evident tension between the eco-

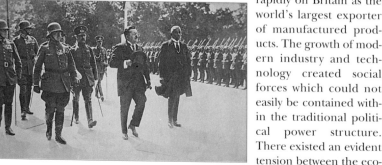

Friedrich Ebert, the Social-Democrat leader who became the first President of the Weimar Republic, is seen here reviewing German troops. The survival of the new parliamentary order in 1919 owed a good deal to the so-called Ebert-Groener Pact in which social-democrats looked to the army to prevent a more radical revolution.

A German 1,000 mark banknote overprinted with one million marks. Hyper-inflation in 1922–23 left many German savers with worthless cash. By November 1923 it was possible to operate with one of hundreds of unofficial currency schemes set up by firms or municipalities to avoid using the deflated mark.

nomic and social modernity of the new Germany and the unre-formed political system.

The new Germany was difficult to accommodate in the international power structure too. A relative newcomer to the club of great powers in Europe, Germany possessed a growing economy, large armed forces (Germany spent more on the military than any other major state in 1914), and very unclear ambitions about what her role in Europe or the wider world might be. Under Bismarck until 1890 the Empire collaborated with the other powers on issues of overseas expansion and Germany won a number of colonies in Africa and the Pacific. Bismarck's main priority was to maintain a balance of power that could guarantee German security in the heart of Europe. In 1890 Bismarck was removed from office by the young Emperor Wilhelm II. He had ambitions to make Germany a greater force in world affairs. Under his rule Germany intervened in the Far East, the Middle East, the Balkans and in Africa. He succeeded in a dozen years in undoing Bismarck's work. Germany found herself isolated diplomatically. Her only firm ally was Austria-Hungary. Wilhelm had no clear blueprint for German foreign policy, which was marked by confusion, hesitancy and a profound sense of insecurity. All of these elements were present in the crisis in 1914 which led to world war. What should have remained a limited conflict in the Balkans between Austrian imperialism and Serbian nationalism became the trigger for a much wider crisis. German support for Austria created a situation from which it was difficult to withdraw once the Balkan crisis escalated. In August 1914 Germany was at war with Russia, Britain and France.

The First World War created the conditions that shaped the later development and ambitions of Hitler's Reich. The conflict generated a great deal of physical hardship on the home front with severe shortages of food and steadily deteriorating social conditions. Organized labour did better than other groups because of shortages of men for the arms factories, but this exacerbated tensions both between and within classes. By the end of the war four years of privation created a widespread hostility which erupted at the very end of the war in a wave of revolutionary violence. In November 1918 Germany was transformed into a democracy and a republic, led by the largest political movement in Germany, the Social Democratic Party. Neither the traditional social forces who lost out in 1918, nor the forces of the popular nationalist right, liked the new system, which they identified with the triumph of Marxism, if of a more moderate variety than in Soviet Russia. The political transition in 1918–19 created a sharp polarization in German society which existed down to 1933.

The infant Nazi Party celebrates German Day in Nuremberg September 1923. The city later became the permanent site of Party annual congresses and Hitler planned a gigantic building programme for the area.

The First World War also ended Germany's long period of economic and trade growth and her pretensions to great power status. German defeat in 1918 left the German economy with a war debt far beyond anything the government could afford to repay. In addition German economic resources were seized by the victorious allies, and a bill of 132 billion gold marks presented to Germany in 1921 as the final schedule of reparation payment. The immediate result was a severe inflation which in 1923, following French and Belgian occupation of the Ruhr to extract reparation in kind, led to the complete collapse of the German currency. The long-term effect was a weakening of Germany's trade position and widespread poverty in a country which had enjoyed buoyant growth for more than forty years.

The new Reich President in 1925, Field Marshal Paul von Hindenburg (left). He was the hero of the Battle of Tannenberg in 1914, when Russian forces were turned back in the early days of the First World War. Hindenburg was seen as a symbol of the old conservative order, but in 1932 social-democrats and Catholics threw their support behind him to defeat Hitler in the Presidential election.

Defeat left Germany internationally isolated and disarmed. Under the terms of the Versailles Settlement drafted in 1919 Germany was allowed to keep only a tiny army capable of keeping internal order (even that was in doubt), and she was excluded from the new international body, the League of Nations, set up at Geneva in 1920. Germany was stripped of territory in Prussia and Silesia, and lost Alsace Lorraine which Bismarck's Reich had taken in 1871. Furthermore Germany was forced to accept her war guilt in the 1919 Treaty. The intense nationalism of Hitler's Reich is easier to understand against a background of economic stagnation and international discrimination which had no parallel in the history of the modern states' system.

The post-war years of the new republic were punctuated by political crisis sparked by the extreme left, which hoped for a more thorough social revolution, and the extreme right, which could not be reconciled to Germany's defeat and subsequent powerlessness. The German Communist Party, formed in early January 1919, was at the centre of the radical revolutionary movement in 1919, and of renewed waves of urban revolt in 1921 and October 1923. The popular nationalist movement gave rise to a number of coup attempts, including the *Putsch* of March 1920 led by Wilhelm Kapp, founder of the right-wing *Vaterlandspartei* during the war, and a smaller coup mounted in Munich in November 1923 by Adolf Hitler's infant Nationalist Socialist Party. The actions of right and left were backed by armed militia and involved considerable bloodshed. The political survival of the Republic rested on the willingness of the tiny armed forces to back the democratic order. The army leaders did so in order to prevent the emergence of a more radical and destructive political system. In 1924 the republic had survived five years of economic crisis and political violence. Helped by the Allied powers, which co-operated with German officials to refound the currency in late 1923, and provided large loans to help expand German economic activity, the system entered a brief period of relative calm. Between 1924 and 1928 there was a period of sharp economic growth and a recovery of trade. The revival benefited large-scale industry and the unionized workforce more than it helped small businesses, artisans or peasants. German society remained polarised, with widespread resentment at the alleged corporate power of big business and big unions. Nonetheless, political conflict became much less violent. Inspired by the right-wing politician Gustav Stresemann, who was briefly Chancellor in 1923 and then Reich foreign minister from 1924 to his death in 1929, many Germans tried to reconcile themselves both with the parliamentary system and with Germany's weak international position. In 1926 Germany was allowed to enter the League of Nations. By the end of the decade German statesmen had succeeded in reducing Germany's sense of isolation. In August 1928 Stresemann signed the Briand-Kellogg Pact of Paris outlawing resort to aggressive war. More significantly in May 1929 he succeeded in renegotiating the reparations issue. The Young Plan, named after the American chairman of the Reparations Commission, Owen D. Young, reduced the overall reparations burden, and promised to remove the remaining occupation forces which had been stationed in the Rhineland since 1919.

It was at just the point that the German economy and German politics were at last coming to terms with the consequences of defeat and post-war crisis that the republican structure was faced with economic disaster. There were signs of impending recession in late 1927 and early 1928. By the spring of 1929, when the Young Plan was agreed, Germany was already in deep crisis with over two million unemployed and collapsing confidence among investors. With the crash in the USA in October 1929 the German economy experienced a catastrophe. Trade fell by two-thirds, the income of farmers and artisans was halved, by 1932 industrial production had fallen by almost half. Even those in work were put on short time. Germans from all walks of life faced two or three years of progressive economic hardship.

There was a powerful sense in German society that having tried to work within the system imposed by the Allies, Germany had been once again the victim of the international order. Confidence in parliamentary government evaporated and the anti-parliamentary right and left became powerful electoral forces. By 1932 the radicalization of the population produced again high levels of political violence. The German Communist Party doubled its parliamentary vote between 1928 and 1932. Fear of Communism pushed many conservative Germans away from the centre-right parties, particularly Stresemann's German People's Party (DVP) and the German Nationalist Party (DNVP), led by the press baron Alfred Hugenberg. Some gravitated to small fringe parties with a strong peasant or small-business bias but the overwhelming bulk moved to support Hitler's National Socialists.

Adolf Hitler salutes the crowds from the balcony of the Reich Chancellery on the evening of 30 January 1933, following his appointment as German Chancellor. An Austrian from a Catholic background, who only obtained full German citizenship in 1932, Hitler won mass support among the largely Protestant communities of northern and central Germany.

The German slump created the breeding ground for Hitler's brand of populism and nationalism, but it did not directly cause Nazi success. This came about partly because of the nature of Hitler's image and appeal — untainted by association with conventional Weimar politics, a man of the people with an apparently profound sense of his historic mission to renew the German nation. But it was also a product of hard electioneering. The Nazi party was a noisy, active, visible movement. It concentrated efforts on local organization and propaganda, tailoring its message to local circumstances, but at the same time promising to bring local problems to a national platform. The ability to win grass-roots support for what many came to see as a crusade for Germany's future turned the Nazi Party into the largest single party in 1932, with the support of more than a third of voters from across the social spectrum.

This success did not necessarily mean power. The lack of a clear parliamentary majority left Hitler in a difficult position in 1932. In May he contested and lost the presidential election. By the end of the year, when the Party itself began to divide over issues of tactics, Hitler's prospect of the Chancellorship looked less likely. He was saved by the revival of the traditional conservative élite and their search for a broader political power base. With the collapse of an effective parliamentary system during the slump, the conservatives rallied in the hope of turning the clock back to the period of élite power before 1914. The chief conservative spokesman, Franz von Papen, who was Chancellor from June to November 1932, and who recognized the need to give old-fashioned conservatism a modern mass following, finally brokered an agreement in late January 1933 in which Hitler would become Chancellor, with a coalition of conservatives and nationalists. On 30 January 1933 Hitler was appointed Chancellor.

Weimar Germany

German democracy, created in defeat in 1919, was faced throughout its existence with popular forces in German society hostile to the parliamentary system.

"And present-day Germany? Fight between democracy and dictatorship, between Jew and German."
Adolf Hitler,
notes for a
speech, 1920

The German Empire reached its fullest extent on 3 March 1918 when the Treaty of Brest-Litovsk was signed with revolutionary Russia. Germany gained control of the Baltic States, Russian Poland, Finland, Ukraine and Georgia, and access to one-third of Russian industry and three-quarters of Russian coal and iron mines. But the new resources of Russia could not be turned quickly enough into resources for war and in a bare eight months Germany was forced to seek an armistice with the western Allies.

By September 1918 Germany's military leaders realized that they faced defeat and sought a way to end the war. They paved the way for the democratization of the country by handing over power to civilian ministers. Growing popular unrest accelerated this trend and on 9 November Friedrich Ebert, leader of the Social Democrat Party, was appointed Chancellor and a republic declared. The German emperor fled to the Netherlands and on 11 November an armistice came into force.

There followed months of revolutionary crisis. An abortive communist rising was staged in January 1919. In Bavaria a Soviet republic was declared, in the Ruhr area workers' militia fought for radical political change until mid-1919. The German revolution was a spontaneous outburst of pent-up anger and frustration. The pre-war radical movement used the crisis of defeat as an opportunity to seize power on behalf of the revolutionary councils that sprang up all over Germany in the winter of 1918-19, but in reality support for a radical change was not widespread. In January 1919 a Constituent Assembly was elected and was dominated by the moderate parties of the centre and left. Meeting at Weimar (hence the name Weimar Republic) the delegates approved a democratic constitution.

The new regime faced enormous difficulties. Domestic order was restored with the help of the army and volunteer forces (*Freikorps*) many of whom were violently hostile to democracy. A crippling peace treaty was signed with

The Berlin cabaret in the 1920s became one of the symbols of what the right-wing saw as decadent modernity.

I/Germany in 1918

- Central Powers
- occupied by Central Powers
- Allies
- neutral countries
- Western front, mid 1918
- Eastern front following the Treaty of Brest-Litovsk, March 1918
- Balkan front, mid 1918
- Italian front, mid 1918
- ...... naval blockade of Germany, 1916-18

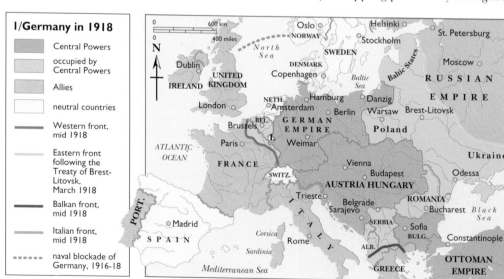

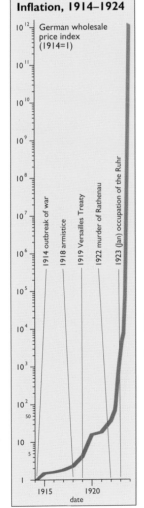

2/Racism in the 1920s

☆ synagogue attack

■ graveyard desecration

the Allies in June 1919. The economy was in deep crisis. A 150 billion marks debt was inherited from the war which weakened the German currency. The low output of food and industrial goods and high state spending to cope with demobilization all fuelled inflation. After a brief period of stabilization in 1921, helped by large loans from abroad, the murder of Foreign Minister Walther Rathenau in June 1922 precipitated a speculative crisis. The mark plunged on world currency markets. When France and Belgium occupied the Ruhr industrial area in January 1923 to enforce reparations the mark collapsed. By November 1923 it was only one-trillionth of its value in 1914 and the state's massive war debt was reduced to only 15 pfennigs. In December 1923 a package was agreed by Germany and the Allied powers to refund the German currency and in 1924 a measure of financial and economic stability was at last introduced, at the cost of impoverishing millions of German savers.

During the following four years a fragile peace returned to German politics. Between 1919 and 1923 the state had been faced with coup attempts from right and left, of which the most serious, the army-backed Kapp Putsch of 1920, was only overturned by a General Strike called in Berlin. German society was bitterly divided. The nationalist right was completely unreconciled to the parliamentary state. They blamed Jews and Marxists for Germany's defeat and the problems of democracy. Anti-semitism became the hallmark of the radical right and led to regular attacks on synagogues and the desecration of Jewish graveyards. Despite the disarmament of Germany enforced through the peace settlement, the army kept alive the skeleton of military organisation, and secretly developed prohibited weapons. Popular militarism continued to flourish, and manifested itself in the growth of para-military organisations tied to the parties of both right and left.

Inflation, 1914–1924

German wholesale price index (1914=1)

- 10^{12}
- 10^{11}
- 10^{10}
- 10^{9}
- 10^{8}
- 10^{7}
- 10^{6}
- 10^{5}
- 10^{4}
- 10^{3}
- 10^{2} / 50
- 10 / 5
- 1

1914 outbreak of war
1918 armistice
1919 Versailles Treaty
1922 murder of Rathenau
1923 (Jan) occupation of the Ruhr

1915   1920
date

3/Germany in the 1920s

Military districts

■ fortification	▷ division command
■ coastal fortification	VI defence region
▨ army group command	● defence region command
▨ army command	○ garrison

The Versailles Settlement

The peace settlement imposed on Germany in 1919 provoked bitter nationalist resentment which permanently undermined the new Republic.

"We know the power of the hatred that we encounter here."
Count Brockdorff-Rantzau at Versailles, 1919

On 7 May 1919 the German delegation to the Paris Peace Conference led by the German Foreign Minister, Count Brockdorff-Rantzau, was presented with the terms of the settlement. There followed a long and bitter argument between German political leaders about whether to accept what they regarded as an unjust and vindictive peace. Only the view of army leaders that further military resistance was futile turned the tide and forced the government to accept. On 28 June German representatives signed the Treaty of Versailles in the Hall of Mirrors in the Palace of the French kings.

The terms of the Treaty went beyond anything even the most realistic German politicians had expected. Germany lost 13% of its territory: Alsace-Lorraine (taken from France in 1870), Danzig, a strip of territory through East Prussia to form a Polish 'Corridor' to the sea, and areas in Schleswig, Silesia and on the Belgian frontier, where plebiscites were held to determine which state the populations should join. The Saar industrial region was placed under international control, but effectively under French influence. The plebiscites were conducted in 1920 and 1921. In Upper Silesia were extensive coal, iron and steel industries which the new Polish state wanted to control. The plebiscite held in March 1921 gave 707,000 votes in favour of Germany, 479,000 in favour of Poland. The Allied powers suspected that undue pressure had been put on Polish workers to vote for Germany, and divided the region on the basis of majority German or Polish voting. The result was that most of Silesian heavy industry went to Poland. In 1926 the German government presented Poland with a bill for 521 million marks in compensation.

I/Germany after the Peace Treaty

- —— borders in 1914
- —— borders after the 1919 peace treaties
- ▨ territory lost by Germany
- ▨ territory lost by Austria-Hungary
- ▨ territory lost by Russia/USSR
- ▨ occupied and demilitarized
- ▨ under international control until 1935
- ▨ Free City under League of Nations
- ····· area in which Germans could not build or improve fortifications
- —— rivers under international control

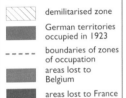

The Treaty also contained punitive economic terms. Germany was forced to pay reparations for war damage, fixed in 1921 at 132 billion gold marks (to prevent Germany paying them with inflation marks). Nine-tenths of the German merchant fleet was confiscated. German rivers were opened to international traffic. Germany's overseas assets, totalling 16 billion marks, were seized. German colonies were taken over by the League of Nations and distributed as mandated territories to Britain, France and Japan.

Finally the Treaty provided for Germany's disarmament to prevent any resurgence of German power in Europe. The army was confined to 100,000 men on long-service contracts to prevent the regular training of conscripts. The General Staff was scrapped, most military installations and training schools closed down, and the Rhineland area demilitarized and occupied by Allied troops. The German air force was abolished, and the German Navy was reduced to a maximum of 6 small battleships of only 10,000 tons each, 6 cruisers, 12 destroyers and no submarines. Any overtly 'offensive' armament was forbidden. An Allied Control Commission oversaw the physical demolition of Germany's military structure and remained in Germany until 1926 verifying German compliance with the disarmament clauses.

The basis for the Allied punishment of Germany was contained in Clause 231 of the Treaty in which Germany was forced to confess her guilt for causing the war in the first place. This clause alienated opinion in Germany across the political spectrum. Versailles turned Germany into a revisionist state *ipso facto*. Only German powerlessness prevented efforts to overturn the Treaty in the 1920s. The Foreign Minister from 1924 to 1929, Gustav Stresemann, argued that Germany would gain more by working within the framework imposed by the Allies than by fighting against it. In 1925 Germany signed the Locarno Treaties with Britain, France, Belgium and Italy recognizing the territorial settlement in the west. In 1926 Germany joined the League of Nations, set up in 1920. German nationalists rejected all compromise with the Versailles system, but the general mood by the late 1920s was of reconciliation from necessity.

2/The Rhineland

▨	demilitarised zone
▓	German territories occupied in 1923
- - -	boundaries of zones of occupation
▓	areas lost to Belgium
▓	areas lost to France
░	area of Saar plebiscite

The major victors in 1918, here represented from the left by Lloyd George (Britain), Orlando (Italy), Clemenceau (France) and Woodrow Wilson (USA), set up a Council in Paris to decide the fate of the defeated states.

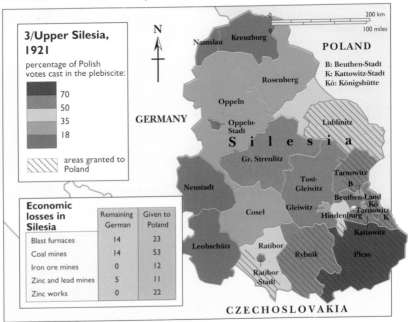

3/Upper Silesia, 1921

percentage of Polish votes cast in the plebiscite:

▓	70
▓	50
░	35
░	18
▨	areas granted to Poland

Economic losses in Silesia	Remaining German	Given to Poland
Blast furnaces	14	23
Coal mines	14	53
Iron ore mines	0	12
Zinc and lead mines	5	11
Zinc works	0	22

The German Slump

The economic slump in Germany was a social and political catastrophe, which opened the way to popular radical movements committed to overthrowing the Republic.

"Smite the Fascists wherever you meet them."
Communist Party slogan, 1931

The effort to restore economic and political stability in Germany between 1924 and 1928 was destroyed by the economic recession between 1928 and 1932. For many Germans the slump was the final straw after defeat, revolution and inflation. Parliamentary politics was undermined and the moderate centre-ground collapsed.

The German economy was fragile throughout the 1920s. The inflation left it short of capital and heavily reliant on large loans from abroad, German trade grew very slowly and home demand was hit by the loss of savings in 1923, the weak state of German agriculture and the cuts in government spending imposed after 1924. When the first signs of recession appeared in 1928 confidence soon evaporated. Well before the Wall Street Crash marked

German printworkers turn out thousands of inflation banknotes during the currency crisis of 1923. Inflation weakened the German economy for the rest of the decade.

1/Germany's foreign indebtness, 1931

512 cumulative debt, 1924–31, in million Reichmarks

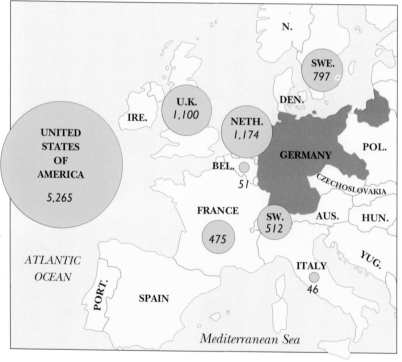

Farm cash income, wages and industrial production in Germany, 1924–36 (index 100 = 1927)

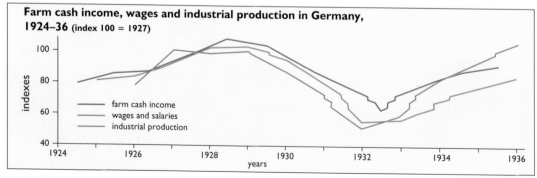

- farm cash income
- wages and salaries
- industrial production

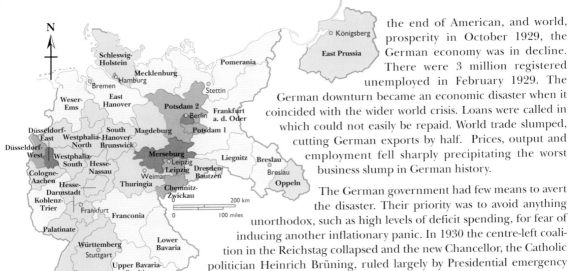

N

2/Communist party support, 1928

percentage of communist vote at elections:

30
22
14
6.5
2

the end of American, and world, prosperity in October 1929, the German economy was in decline. There were 3 million registered unemployed in February 1929. The German downturn became an economic disaster when it coincided with the wider world crisis. Loans were called in which could not easily be repaid. World trade slumped, cutting German exports by half. Prices, output and employment fell sharply precipitating the worst business slump in German history.

The German government had few means to avert the disaster. Their priority was to avoid anything unorthodox, such as high levels of deficit spending, for fear of inducing another inflationary panic. In 1930 the centre-left coalition in the Reichstag collapsed and the new Chancellor, the Catholic politician Heinrich Brüning, ruled largely by Presidential emergency decree. Parliamentary government became almost superfluous and public confidence in democracy evaporated. In the 1930 Reichstag election the Nazi Party made its first real electoral breakthrough, but the Communist Party, formed in 1919 during the revolutionary crisis, began to attract the votes of urban workers disillusioned with social-democracy and the failure of German capitalism. In the major cities the radical left and right fought running street battles which left hundreds dead. By early 1932 Communist membership rose to 287,000. It was particularly strong in Saxony, Berlin and the northern port cities. In June 1932 the party set up an Anti-Fascist Action Front to organize a concerted fight against Nazism, but its continued hostility to the SPD, which was regarded as a reactionary 'social fascist' party, weakened its appeal. The revival of the revolutionary threat from the left did more than anything during the slump to push centre and right-wing voters towards the nationalist extremes. The centre-right parties collapsed electorally. With more than eight million people out of work in 1932 moderate solutions no longer appealed.

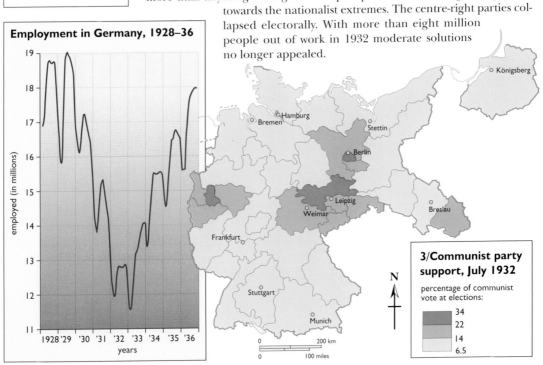

Employment in Germany, 1928–36

employed (in millions)

1928 '29 '30 '31 '32 '33 '34 '35 '36
years

N

3/Communist party support, July 1932

percentage of communist vote at elections:

34
22
14
6.5

The Nazi Party before 1933

From small beginnings Hitler turned the Nazi Party from a fringe movement of a few hundred people into a mass party fighting for parliamentary power.

"Barely 16, I joined the Party and fought and propagandized eagerly for the new view of the world..."
SA recruit, 1934

The roots of the NSDAP (Nazi Party) lie in the southern German province of Bavaria. In September 1919 Adolf Hitler became member number 55 of a tiny German Workers' Party founded by Anton Drexler, a toolmaker. In February 1920 the party adopted a 25-Point Programme drafted by Hitler and the economist Gottfried Feder which had clear anti-capitalist sentiments and was strongly anti-semitic. In April 1920 the party changed its name to National Socialist German Workers' Party, and in July 1921 Hitler, who had become a local celebrity with the power of his public speaking, was elected its chairman. Two years later, at the height of the inflation crisis in November 1923, Hitler launched an armed coup in Munich which was crushed by the local police.

For his part in the attempted coup, Hitler was sentenced to five years' imprisonment in Landsberg prison. He served only nine months, but during that time wrote the first volume of *Mein Kampf* (My Struggle) which became the bible of the new movement re-founded by Hitler at a Congress in Bamberg in February 1926. With the help of Gregor and Otto Strasser, and the Party's young propaganda expert, Joseph Goebbels, the organization became firmly established in northern Germany where it attracted increasing support. In the early 1920s the Party's socialist and populist rhetoric gave it a strong working-class base, but by the late 1920s many more petty-bourgeois and white-collar workers joined the movement. By the early 1930s approximately 60% of the members came from middle-class backgrounds, while the remaining

Hitler the dreamer: a publicity photograph of the Führer taken during his nine-month imprisonment in the fortress of Landsberg in 1924 following the Munich Putsch.

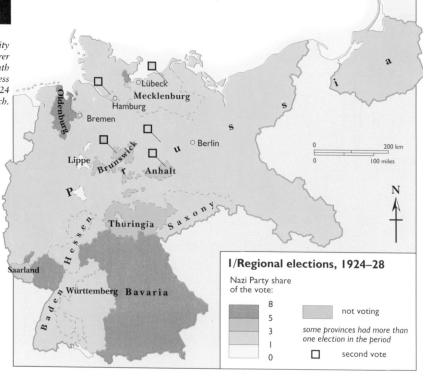

Lübeck
Mecklenburg
Hamburg
Oldenburg
Bremen
Lippe
Brunswick
Berlin
Anhalt
Thuringia
Saxony
Hessen
Saarland
Baden
Württemberg
Bavaria

1/Regional elections, 1924–28

Nazi Party share of the vote:

8	
5	
3	not voting
1	
0	*some provinces had more than one election in the period*

□ second vote

0 — 200 km
0 — 100 miles

N

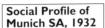

Small businesses 2.1%
Farmers 0.4%
Others 8.5%
Unskilled workers 7%
Students 12%
Professionals 4.3%
White collar 20.5%
Skilled workers 45.1%

Nazi Party Congresses, 1923–33

First Party Day, Munich January 27–28, 1923

German Day Rally, Nuremberg September 1, 1923

Refounding Congress, Weimar July 4, 1926

Day of Awakening, Nuremberg August 20, 1927

Party Day of Composure, Nuremberg August 2 1929

Party Day of Victory, Nuremberg September 1 1933

40% of manual workers contained a large number of craft or rural workers who had not been organized in the socialist trade union movement. The Party's para-military wing, the *Sturmabteilung* (SA), founded in November 1921, was more heavily proletarian in character. Prior to 1933 approximately 56% came from working-class backgrounds, a figure that rose to two-thirds in 1933–1934.

After Hitler's release from prison he committed the movement to electoral politics, insisting that power should be attained through legal procedures. The Party made little progress between 1924 and 1928, although it had more success in local elections where the Party was well-organised. The breakthrough came with local elections in Thuringia and Saxony in 1930 when the Party at last won more than 10% of votes cast. The success owed something to the emphasis placed by the Party on local political activity. Local branches and cells were set up all over Germany and the Party deliberately sought publicity, good or bad.

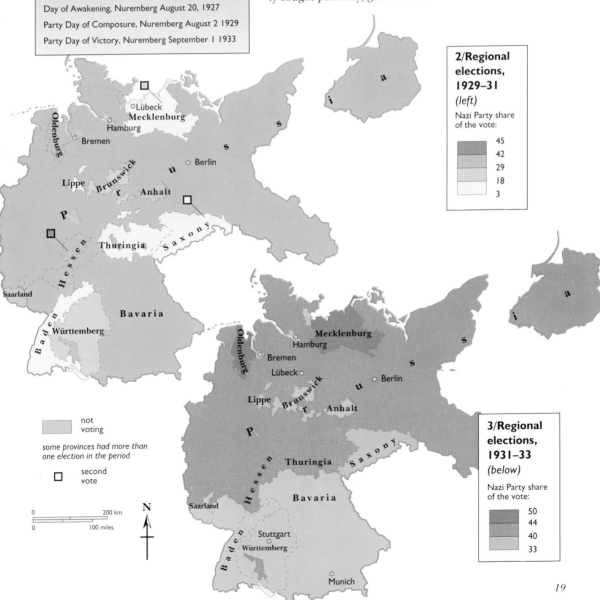

2/Regional elections, 1929–31 (left)

Nazi Party share of the vote:
- 45
- 42
- 29
- 18
- 3

3/Regional elections, 1931–33 (below)

Nazi Party share of the vote:
- 50
- 44
- 40
- 33

not voting

some provinces had more than one election in the period

□ second vote

0   200 km
0   100 miles

N

The Legal Path to Power

Though publicly hostile to the parliamentary system, Hitler insisted that the Party should achieve power through the ballot box.

"For us Parliament is not an end in itself, but merely a means to an end."
Adolf Hitler,
Munich, 1930

In the Reichstag election results of 20 May 1928 Nazi hopes of an electoral breakthrough were disappointed. The Party polled 100,000 fewer votes than in 1924, and in the areas where the greatest organizational effort had been made, in the major cities, the results were trivial; in Berlin 1.4% of the votes, in Hamburg 2.6%, in the Ruhr area 1.3%. With only 12 deputies the Party was one of a number of small splinter parties with a radical nationalist outlook.

In 1928 it seemed that Hitler's aim of achieving power through legal means had failed. The only positive result was the support shown in a number of northern and central regions which were predominantly rural, and in which the Party had made only a limited electoral effort. It was here that the Party began to recognize its true constituency and over the next four years it concentrated on mobilizing the support of the conservative masses—

1/Religion in Baden

- Catholic majority
- Protestant majority
- balanced districts

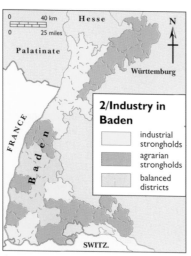

2/Industry in Baden

- industrial strongholds
- agrarian strongholds
- balanced districts

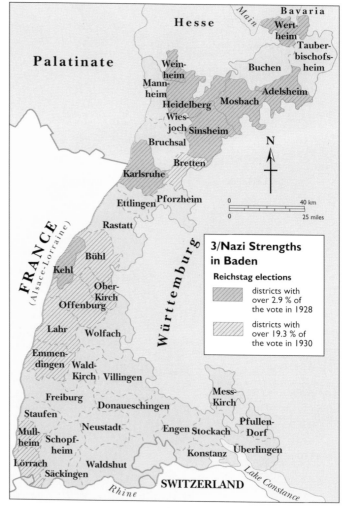

3/Nazi Strengths in Baden

Reichstag elections

- districts with over 2.9 % of the vote in 1928
- districts with over 19.3 % of the vote in 1930

The Party headquarters building in Munich, known as the 'Brown House'. It was here that the Nazi electoral campaigns were master-minded.

peasants, craftsmen, small businessmen, officials—who shared the Nazis' dislike of Marxism and were politically alienated from the modern urban-based democracy. Walther Darré, as Reich organizer for rural areas in 1930, smothered the countryside with party cells and party propaganda. In the largely protestant areas of northern and central Germany the Nazis picked up a large rural protest vote. In the September 1930 election they gained 103 seats and became at last a more formidable electoral threat.

Nazi success rode on the back of the recession, but it was the product of an intense campaign of organization, fund-raising and propaganda. Party leaders targeted national organizations of peasants, white-collar workers and craftsmen with the object of winning support for the movement. By 1932 it was a mass party, winning a larger share of the popular vote than any party since 1919. A second election, in November 1932, brought a fall in Nazi votes. But as the largest party it could not be ignored and in January 1933 the conservative politicians around the President, von Hindenburg, invited Hitler to take the Chancellorship in the hope of using his mass following to underpin a conservative remodelling of the Weimar state. when Hitler accepted the office on 30 January he had at last gained power by legal means.

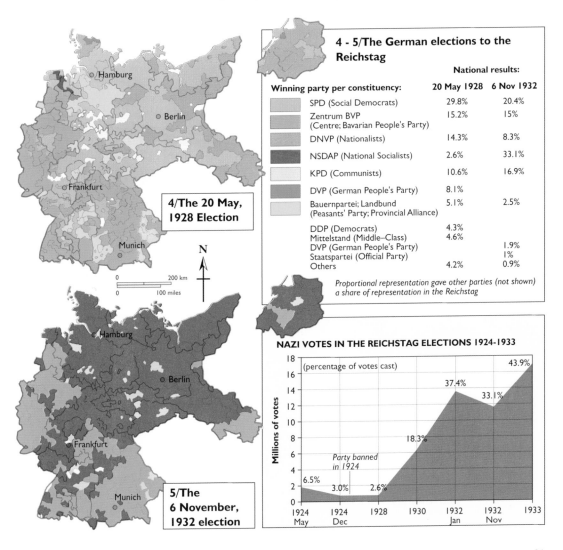

4 - 5/The German elections to the Reichstag

Winning party per constituency:	National results:	
	20 May 1928	6 Nov 1932
SPD (Social Democrats)	29.8%	20.4%
Zentrum BVP (Centre; Bavarian People's Party)	15.2%	15%
DNVP (Nationalists)	14.3%	8.3%
NSDAP (National Socialists)	2.6%	33.1%
KPD (Communists)	10.6%	16.9%
DVP (German People's Party)	8.1%	
Bauernpartei; Landbund (Peasants' Party; Provincial Alliance)	5.1%	2.5%
DDP (Democrats)	4.3%	
Mittelstand (Middle–Class)	4.6%	
DVP (German People's Party)		1.9%
Staatspartei (Official Party)		1%
Others	4.2%	0.9%

Proportional representation gave other parties (not shown) a share of representation in the Reichstag

4/The 20 May, 1928 Election

Hamburg
Berlin
Frankfurt
Munich

N

0   200 km
0   100 miles

5/The 6 November, 1932 election

Hamburg
Berlin
Frankfurt
Munich

NAZI VOTES IN THE REICHSTAG ELECTIONS 1924-1933

(percentage of votes cast)

43.9%
37.4%
33.1%
18.3%
6.5%
3.0%
2.6%

Party banned in 1924

Millions of votes

| 1924 May | 1924 Dec | 1928 | 1930 | 1932 Jan | 1932 Nov | 1933 |

II: Establishing the Dictatorship

In January 1933 Hitler's appointment as chancellor opened the door to the Nazi Party's 'brown revolution'.

"When we came to power in 1933 the German people found itself in the midst of a mighty historical transformation."
Adolf Hitler, May 1936

The 1934 poster 'The New Germany' deliberately links the Nazi stormtrooper with the soldiers of the First World War. The Nazi Party gained from identifying its cause with German militarism. Nazi language and ritual relied very largely on a militaristic idiom, which other parties failed to exploit.

When Hitler came to power in January 1933 many in the Nazi Party expected a revolution. They had been promised a new Germany and they expected Hitler to redeem that promise. Hitler's position was in reality more precarious. He was the head of a coalition government, composed mainly of more traditional conservatives. Only two other Nazis were in the cabinet, Wilhelm Frick as Minister of the Interior and Hermann Göring as Minister without Portfolio. He needed the support of other parties to win a majority in the Reichstag. To try to strengthen his position he called fresh elections, which were held on 5 March. The Nazi Party won 43.9% of the vote, winning for the first time a much larger share of the vote in the Catholic south. This was still short of a majority. On 23 March Hitler presented an Enabling Act to the Reichstag which was designed to give him permanent emergency powers. Careful wooing of the Catholic Church produced support from sufficient Centre Party deputies to pass the bill. Hitler was freed from the limitations imposed by a parliamentary system and could embark on the construction of a dictatorship.

There was no agreement even among Party members about what precisely a National Socialist state would look like, but the ideological fixation with Germany's national and moral renewal, the war on Marxism and enemies of Germandom and an end to class conflict and party politics was clear enough. What followed in 1933 was a process known as *Gleichschaltung* or 'co-ordination'. Party enthusiasts and the SA had already begun an informal and often violent programme of their own from January 1933 in which they intimidated or murdered opponents and pressurized businesses and institutions into accepting Party orders. The wave of lawlessness provoked widespread disapproval, but Hitler allowed it to continue until the summer, interfering on occasion where the violence threatened to get out of hand. Backed by the Enabling Act Hitler was able to carry out a more formal and systematic *Gleichschaltung* from above. The first victims were the political parties. The KPD was effectively banned in late February 1933 following the burning of the Reichstag building, which Hitler blamed on a Communist conspiracy. The ban was given a retrospective legality on 26 May 1933 with the Law for the Seizure of Communist Assets. The Trade Unions were taken over on 2 May and their assets seized, and the Social-Democratic Party was banned on 22 June as an organization 'hostile to the state'. The other right-wing parties represented in the original parliamentary coalition wound themselves up and some members transferred to the Nazi Party. The Centre Party, anxious to defend Catholic interests, and deeply split on its attitude to Hitler, survived until 5 July when it formally dissolved its organization. On 14 July the Law against the Establishment of Parties was promulgated and Germany became a one-party state. A Law on Plebiscites announced on the same day established the referendum as the instrument for popular approval of the government's actions. The first plebiscite on 12 November 1933 produced 89.9% in favour of Hitler's deci-

A fanfare of Hitler Youth at the Party Rally in Nuremberg, 1935. The rallies were the centrepiece of the Party calendar, at which the leadership was expected to indicate future policy and ideological development. The 1935 rally saw the so-called Nuremberg Laws which launched the introduction of racial inequality.

Hitler in an unusual pose, dressed in the traditional Lederhosen. Hitler always took great care with his appearance in public. Only very rarely was he photographed wearing glasses, or eating and drinking. He seldom smiled for the camera.

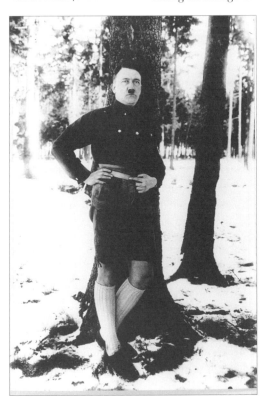

sion to withdraw Germany from the League of Nations. Subsequent plebiscites in 1934 and 1938 produced the same overwhelming support for the regime.

Co-ordination was enforced throughout German society. No institution was immune. A Concordat with the Catholic Church signed on 8 July 1933 in Rome preserved the confessional interests of the Catholic Church but forced Catholics to abandon politics to the Nazi Party. All areas of social life from youth groups to choral societies were brought under Party influence as far as possible. The Party apparatus acted as a shadow state, bullying and harassing any individual or association hostile to Party interests. The cost of continued resistance was violence against person or property, or in many cases arrest and imprisonment in one of a number of camps for political dissidents set up and run by Party thugs from early 1933. The 'Partification' of German life after 1933 helped to create an illusion of popular consensus even in areas where support for Hitler had been much more subdued before 1933.

Party activists could be forgiven for thinking that some kind of revolution had occurred in 1933. Hitler found himself caught between radical elements in the Nazi movement (especially in the SA under Ernst Röhm) who called for a more thorough shake-up of German society, and his conservative allies who dominated army, foreign office and the economy, and who were anxious to avoid a more fundamental revolution. In the end Hitler came down on the side of his conservative allies, partly from fear that the revolutionary elements in the movement might undermine his own position as Chancellor and Party Leader, or dispense with his services altogether. In late June 1934, taking advantage of a gathering of SA leaders at Munich, Hitler's SS bodyguard, armed by the regular army, murdered Röhm along with other prominent supporters of the Nazi left and a number of former political opponents.

The termination of the revolutionary threat brought Hitler the support of the army and other conservative elements. In August 1934, following the death of the President, von Hindenburg, Hitler merged the office of Chancellor and President together, and on 2 August the armed forces were required to swear on oath of loyalty to Hitler. In September the conservative banker, Hjalmar Schacht, was made Economics Minister, a clear sign that Hitler did not want any radical overhaul of the German economic system. Since 1933 businessmen had been uncertain about how far the anti-capitalist rhetoric of the Party might be used to attack German capitalism. Hitler's view of the economy was essentially pragmatic. He needed an economic recovery in order to stabilize German society and end unemployment. He pursued orthodox economics to restore confidence in the business community, and he recruited conservatives like Schacht, or the iron and steel industrialist Gustav Krupp, who was head of the Reich Industrial Association in 1933, in order to keep business as part of the conservative alliance created in the early months of the regime.

Hitler's own SS guard, parading on 9 November 1935, the twelfth anniversary of the failed Munich Putsch. Hitler's first ignominious entry into national politics later became a major date in the Nazi calendar of festivals, which included Hitler's birthday (20 April) and the annual Party Day (1 September).

Below: An advertisement for the 1936 Olympiad in Berlin, held in the first two weeks of August. There were strong protests in the USA against participation, but following German assurances that Jewish competitors would not be discriminated against (which was patently not so) a boycott of the Games was avoided.

The establishment of comprehensive wage controls and the termination of free collective bargaining for labour, both incorporated in the Law for the Organization of National Labour of 16 January 1934, ended the conflicts between labour and capital that had punctuated the troubled history of the Weimar Republic.

The honeymoon with the conservatives lasted only two years. With the end of the Röhm threat Hitler could control the pace of change more effectively, but there was no real pause in the process of co-ordination. The German *Länder* (provinces) which had enjoyed an independent political existence, with their own legislatures and administration, were brought directly under central Reich control in the Law for the Reconstruction of the Reich published on 30 January 1934. The Reich justice system was unified in April 1935, and the police and the security services were brought under Himmler's control in June 1936. The precise relationship between the Nazi Party organization of regional Gaus, run by Party Gauleiter, and the new centralized state administration was never satisfactorily resolved, and state/party relations remained a source of perennial friction throughout the life of the Third Reich.

In 1937–8 Hitler began to loosen and then cut his ties with the German conservatives. The decision to adopt a strategy of autarky, or self-sufficiency, for the economy in autumn 1936, and to accelerate the pace of rearmament (see pp. 46–9) brought him into conflict with the military and the business community. His revelation in November 1937 that he was planning small wars in the near future alienated diplomats and soldiers who were anxious not to run risks. Rearmament and economic development were brought under the wing of Hermann Göring, a close Hitler confidant. Schacht was forced to resign in November 1937; the War Minister, von Blomberg, was eased out in a staged scandal over his private life; the Foreign Minister von Neurath was sacked in favour of the Party foreign affairs spokesman, Joachim von Ribbentrop. The only conservative to retain Hitler's favour was the colourless Finance Minister, Count Schwerin von Krosigk, who kept his job throughout the 12 years of Nazi rule.

The political changes of 1936–8 opened the way to a more fundamental change in the character of the regime. State control over the economy was strengthened in every area, leaving German businessmen with little independence by 1939. The SS, holding a monopoly on security in the Reich, became a major power centre. The military, following Hitler's assumption of an active Supreme Command (OKW) in February 1938, became entirely dominated by Hitler except in areas where their technical expertise continued to give them some room for manoeuvre. Hitler assumed a much more personal dictatorship in the late 1930s, dispensing with committee work, fabricating a 'Führer myth' of his own infallibility and competence, posing as a truly national figure, set aside from his Party. His personal popularity, so far as it can be judged within an authoritarian system, seems to have reached its height in the years

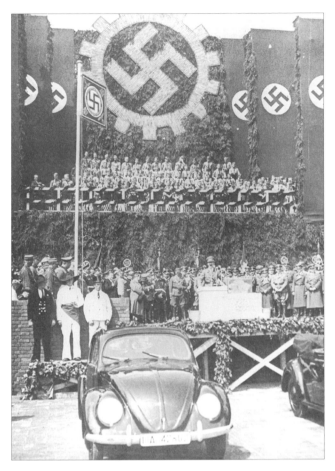

Hitler launches the Volkswagen (People's Car) in 1939. He was a great enthusiast for mass motoring, believing that it would help to bind Germans together and give them a proper sense of the diversity and character of the German space. The car was designed by Ferdinand Porsche, who produced high-performance sports cars. Only a handful were produced before the war, and a military version was built in its place. The car went on to become one of the major success stories of the post-war German economic miracle.

1938 to 1940. A Nazi consensus was forged to replace the broader conservative-nationalist coalition which had existed in 1933.

The consensus was to some extent an obvious fiction. There were those in German society who were never reconciled to the movement but who lacked the courage or opportunity to dissent. They engaged in what was called an 'inner emigration', paying lip-service to the regime but denying it in their private thoughts. Elements remained who were willing to run great risks in promoting political opposition through the dissemination of illegal literature or the holding of clandestine party meetings. Other aspects of dissent hardly amounted to political resistance, though the regime often defined them as such, but were the product of a genuine popular dislike of one or another aspect of Nazi policy. The two most well-known examples were the outcry in Bavaria against the decision to remove crucifixes from schools, and the Church-led condemnation of the 'euthanasia' programme for the destruction of the mentally and physically disabled, ordered by Hitler in 1939. There were other examples. None constituted a serious political threat to the survival of the dictatorship, and it is significant that in both these two cases the regime publicly altered its position.

Political opposition was dangerous and the penalties severe. By the late 1930s it was of little significance. It survived in the most unlikely quarter, among the senior officers around Hitler, who toyed with the idea of a coup in 1938 and again in 1939, but hesitated to do the deed because of Hitler's evident popularity. The most notorious aspect of the Third Reich, its persecution of the Jews, provoked surprisingly little resistance. Beginning with Party anti-semitic outrages in 1933, racism became a characteristic feature of the regime. Based on a crude racial pseudo-science the Party, or more particularly a smaller group of committed racists within it, pursued the goal of racial hygiene — the building of a pure German stock unsullied by the hereditarily ill and disabled, by Jews, gypsies and other 'alien' peoples. The Jews were singled out as the greatest threat to Germany's racial survival, and they were the victims of official discrimination. Their property was confiscated, welfare payments suspended, their jobs terminated and professions boycotted. Hundreds of thousands emigrated after forfeiting their possessions to the Reich. Jewishness was defined as the antithesis of all things German. Hitler saw the Third Reich as the saviour of German blood and German culture. The real revolution in 1933 was a revolution of values, not of classes.

The Organization of Authority

During 1933 and 1934 the Weimar system was replaced by a one-man, one-party dictatorship, dominated by the principle of 'leadership'.

"Our Constitution is the Will of the Führer."
Hans Frank, May 1936

The overthrow of the parliamentary system after Hitler's appointment as Chancellor in January 1933 was achieved gradually. The first step came with the so-called Reichstag Fire decree which was published following the burning down of the parliament building by a young Communist. The decree of 27 February 1933 gave the government emergency powers which were used against the Communist Party in particular. Hitler called fresh elections for March and although these did not give the Party an overall majority, enough non-socialist deputies were persuaded to vote with the Nazi Party for an Enabling Act on 24 March 1933, which gave Hitler dictatorial powers.

Over the summer months all other political parties were dissolved. The Nazi party could now transform the constitutional structure. Party leaders were appointed in the provinces to the post of Reichsstatthalter, designed to bridge the gap between party and state, with wide executive powers. The civil service was purged of elements deemed to be politically or racially undesirable. Approximately one-quarter of Prussian officials, and one-tenth of those in the whole of Germany, were removed from office. Hitler kept the structure of Party organization and state administration apart, but on his appointment as 'leader' in August 1934 (which fused the role of President, Chancellor and Party leader in one office) he stood at the point of the whole pyramid of state and party organization.

I/The German Länder, 1933-35

- ■ S.A. group headquarters
- ■ Reichstatthalter

At the same time the German political structure was strongly centralized. In February 1934 the interior Minister, Wilhelm Frick, took over the ill-defined legislative functions of the Reichsstatthalter, and on 30 January 1935 the provincial parliaments and state authorities were subordinated directly to Berlin, ending the federal

The radical Hitler and the reactionary Hindenburg side by side on a poster in 1933. When Hindenburg died in August 1934 Hitler merged the offices of President and Chancellor and became the German leader 'Fuhrer'.

The Führer
(President, Chancellor and Party Leader)

STATE		PARTY
Reich Chancellory		Party Chancellory
Reich Minister		Reichsleiter
State Secretary	Reichsstatthalter	
Minister President		Gauleiter
District President		
Landrat (local executive)		Kreisleiter
Bürgermeister (mayor)		Ortsgruppenleiter
State Officials (Beamten)		Politische Leiter (political leaders)

character of the German polity first set up by Bismarck in 1871. Police and security powers were also centralized under Heinrich Himmler, head of the SS. A number of corporate bodies were established—the Labour Front, the Reich Food Estate, the Reich Chamber of Handicrafts etc.—which were designed to replace with one national organization the many different institutions which had flourished in the pre-1933 liberal system.

These changes effectively made both parliament and cabinet redundant. Hitler disliked ruling by committee and soon abandoned regular attendance at cabinet. The loose political coalition which survived the first months of Hitler's rule was gradually eroded as key offices of state were filled with Party lieutenants. This process was completed in February 1938 when Hitler became Supreme Commander of the armed forces, and replaced the war ministry with a Supreme Headquarters (OKW). But by then Hitler was regarded as the source of final authority on all issues, however trivial. This had the effect of slowing up the business of state, since Hitler was an unenthusiastic administrator, and it created an intense and uncontrolled competition between state and party agencies for the attention of Hitler, or some other Nazi patron. Such a system more resembled an eighteenth century absolutist court than a modern bureaucratic state.

Hitler's cabinets, 1933–38

1933	Hitler (Chancellor)	1938
von Papen (Centre)	Vice-Chancellor	post suspended
von Neurath (non-party)	Foreign Minister	Ribbentrop (NSDAP)
Frick (NSDAP)	Interior Minister	Frick (NSDAP)
von Krosigk (non-party)	Finance Minister	von Krosigk (non-party)
von Blomberg (non-party)	War Minister	replaced by Hitler's Supreme HQ
Hugenberg (DNVP)	Economics Minister	Funk (NSDAP)
Göring (NSDAP)	Air Minister	Göring (NSDAP)
Rust (NSDAP)	Education Minister	Rust (NSDAP)
Seldte (Stahlhelm)	Labour Minister	Seldte (Stahlhelm)
von Eltz-Rübenach (non-party)	Transport Minister	Dorpmüller (non-party)
Goebbels (NSDAP)	Propaganda	Goebbels (NSDAP)

The Nazi Party I: A Parallel State

The Nazi Party was organized as a parallel state, with its own leaders, its own chancellery, its own network of regional administrators.

"The Party is the leadership and the legislature. The State is the administration."
Adolf Hitler, July 1936

The Nazi Party saw itself as more than simply another political party. It was presented as a national movement rather than a class or sectional interest, and many Nazi leaders hoped to merge party and state together after the seizure of power in 1933. Hitler as head of state and of the party maintained a dualistic structure, but in practice Party officials were able to exercise considerable influence on public affairs and the SS, the Party élite, became by the war a law unto itself, a virtual state within a state (see pages 100–101).

The Party membership grew rapidly following the Nazi success at the polls in March 1933. An additional 1.6 million members were enrolled from March up to 1 May, when a ban was imposed on further membership. It was still pos-

A painting by Felix Albrecht of Nazi SA men, portrayed as heroes of the movement in the tough street battles of the pre-1933 years. Once in power the SA launched a campaign of violence and harassment against all alleged enemies of the movement.

I/Nazi organization, 1938

- – – – – pre 1938 Gau administrative border
- ⬛ SA Group headquarters
- • Gau capitals (seats of Gauleiters)
- ◻ Party headquarters
- △ Party congress centre
- ● Hitler's retreat

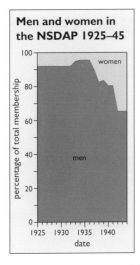

Men and women in the NSDAP 1925–45

Ostland
Königsberg

Pomerania

MEMELLAND

East Prussia

P O L A N D

Silesia
Silesia   Breslau

sible to join the party via the SA or other party organizations and by 1939 membership had reached 5.3 million against 849,000 in 1933. During the war millions more joined, including an increasing number of women and factory workers, both groups under-represented in the movement in the 1930s. By the end of the war there were 8 million Party members, and the proportion of manual workers among the new members grew steadily to reach over 40% in 1944. The same year 35% of new members were women, against a figure of only 5% of new members in 1933.

Many Germans joined the Party because it was expedient to do so or because membership was more or less a condition of particular kinds of employment—the civil service, or in schools. In early 1933 there were 13,000 teachers in the Party, but by May there were 71,000. The social composition of the Party did not entirely reflect its electoral constituency. Peasants provided perhaps as much as 50% of Party support in 1932, but by 1942 constituted only 15% of members. By the war the leadership of the movement was dominated by men in white collar occupations, or from a craft or working-class background, making this Party more socially 'modern' than its reactionary image as a friend of peasants and artisans suggested.

The Party was organized in such a way that it covered the entire geographical area of Germany, and the entire population. The key institution was the Gau, or Party province. The Gauleiter set up a parallel structure to the existing state authorities, staffed by Nazi experts and officials. In areas such as labour supply and industrial investment they could exert considerable pressure on the central authorities. An ambitious and unscrupulous Gauleiter could interfere in local affairs to build his own personal political empire and to protect the interests of his Gau. The SA, led in 1933 by Captain Ernst Röhm, also had a national organization. In 1933 the SA numbered some 425,000. Its leaders were impatient to impose a radical social revolution on Germany after March 1933, and Röhm himself had ambitions to turn the SA into a new national army to challenge the traditional armed forces. Hitler did not want a Party takeover of the state and its functions, and on 30 June 1934 in the so-called Night of the Long Knives, Röhm and many of his SA colleagues were murdered on Hitler's orders. In 1935 the SA was remodelled and many SA men demobilized. It continued to play an important part in the life of ordinary Party members, but the balance of power in the movement tilted to the SS and the army of approximately 400,000 permanent Party officials.

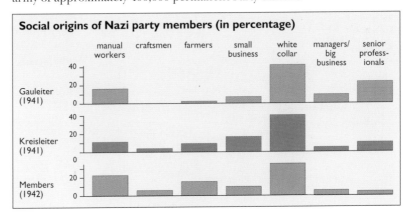

Social origins of Nazi party members (in percentage)

The Nazi Party II: Enforcing Consensus

The Nazi Party found all kinds of ways to enforce loyalty and create consensus, particularly among young people.

"People were always coming and saying why haven't you hung out a flag for Hitler's birthday ... It was very dangerous if you didn't..."
Librarian, 1930s

The Nazi Party established a role in German society that went far beyond the organization and its members. The movement had always emphasized the importance of ritual and display, and its conspicuous activity was one of its strengths in competition with the other parliamentary parties. Once in power a number of small but significant rituals were used to help identify friend from foe. The Hitler salute was one. Another was the requirement that swastika flags or pictures of Hitler should be displayed on Nazi festivals or during official visits was an unsubtle way of enforcing consensus. The costs of non-compliance were not always dangerous, but the social and career advantages of conformity outweighed for many any qualms of conscience.

The Nazi Party targeted German youth in particular. In 1933 there was a proliferation of youth groups reflecting political or confessional interests. By 1936 most had been absorbed into the Hitler Youth, a broad-based movement covering children of all ages and both sexes. First founded in 1926, the Hitler Youth developed into a gigantic movement in the 1930s, with more than 7.7 million members.

I/Hitler Youth organization

- - - - - district border after 1939

district headquarter

Nazi girls' youth organization hiking regulations

Daily goal for a 1 or 1.5 day hike

age	length	speed	size of pack
10–11	10 km	3 km p.h.	sandwich pack
12–14	15 km	4 km p.h.	6 lb pack

Daily goal for a hike of several days

12–14 years old

	length	speed	size of pack
day 1	15 km	4 km p.h.	8 lb pack
day 2	10 km	4 km p.h.	8 lb pack
day 3	10 km	4 km p.h.	8 lb pack
day 4	rest day		

1 hike per month was compulsory

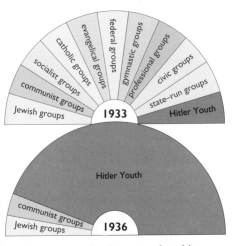

Youth organizations membership

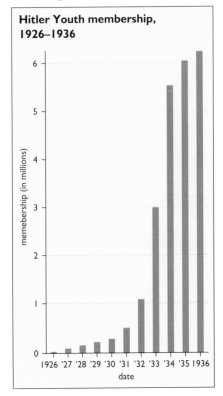

From December 1936 it was made a state agency, under the leadership of Baldur von Schirach, and membership was compulsory. Boys organized in the *Jungvolk* to age 13, and then the Hitler Youth from 14 to 18. Girls joined the *Jungmädel* to age 13, and then the League of German Girls to age 18. Limited political education was combined with sports activities, compulsory cross-country hikes, with strict regulations, and charity work. Boys received a certain amount of para-military training. Both sexes graduated at 18 into some form of labour service, designed to emphasize the classless character of the organization and the nobilty of work.

Women were also targeted by the party. Heavily under-represented in the Party itself, which was permeated by values of aggressive masculinity, the Nazi women's organization emphasized the role of women in the home and as childbearers. The Nazi Women's front (Frauenfront) was set up in 1931. In 1934 it was reorganized under a female Führer, Gertrud Scholtz-Klink, who took reponsibility for women's affairs in all areas of the Party. Radical women's movements were closed down. So too were birth-control clinics and institutes. Nazi propaganda stressed the traditional feminine roles of *Kinder, Kirche, Küche* (children, church and kitchen). In practice many women stayed in paid employment or worked on small farms and businesses. By 1939 over one-third of the German workforce was female, though top jobs in the professions and business were almost exclusively the preserve of men.

Membership of Nazi subsidiary organizations, 1934	
Full time and volunteer workers in the Nazi Party	
Political organizaton	373,000
National-Socialist Factory Organization	120,000
NS Women's Organization	53,000
NS Teacher's Organization	12,700
NS Doctor's Association	1,500
NS Lawyer's Association	1,600
NS Welfare Organization	68,000
Propaganda Offices	14,000
Hitler Youth	205,000
Reich Labour Service	18,500
Total of all NS functionaries	1,017,000

A display of synchronised callisthenics at the Nuremberg Party Rally of 1935. Sport was one of the chief avenues through which young Germans were led into the movement.

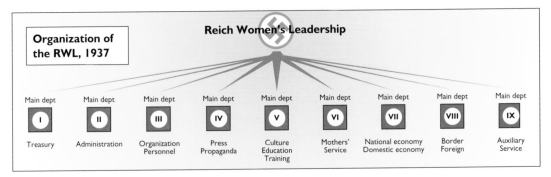

The Economic Revival

Hitler saw Germany's economic revival as the essential task in 1933, on which the political survival of the regime rested.

"Unemployment is the gigantic problem set before us for solution: in face of that problem everything else must take second place!"
Adolf Hitler, 1933

When the Nazi Party came to power in 1933 there were almost nine million fewer people in work than in 1929, two-fifths of the workforce. Unemployment was widely seen to be the key issue, and Hitler saw its eradication as an essential pre-requisite for the achievement of his other military and foreign-policy ambitions.

Hitler was keen to avoid economic experimentation, much to the disappointment of more radical members of the party. He continued and enlarged the work creation programmes inherited from the last Weimar governments. Under the leadership of the man Hitler appointed to head the Reichsbank in March 1933, the conservative Hjalmar Schacht, Germany reduced her dependence on international financial markets by a virtual default on her debts. The banking system and the capital market came under close state supervision to avoid any weakening of the mark and to open up the blocked arteries of credit. Above all the government undertook to stimulate demand by tax concessions and subsidies (particularly to car owners and house owners) and to boost investment by state loans for roadbuilding and repair, public buildings, transportation and military revival. Public spending rose from 15% of Gross National Product (GNP) in 1929 to 33% by 1938. Between 1933 and 1936 some 21 billion marks was invested by the state.

The effects of greater state spending and control were marked. By 1937 registered unemployment was down to just under one million. The pre-slump peak of GNP was exceeded by 1935. The level of industrial production achieved in 1928 was surpassed in 1936. By 1939 the economy had grown 33% above the peak in the late 1920s. Much of this growth benefited heavy industry and construction. Export growth was sluggish in the 1930s, and the trade levels in the 1920s, already lower than 1913, were never recovered. Nor did consumers bene-

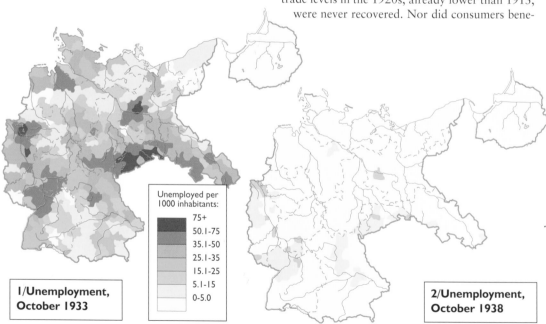

Unemployed per 1000 inhabitants:
- 75+
- 50.1-75
- 35.1-50
- 25.1-35
- 15.1-25
- 5.1-15
- 0-5.0

1/Unemployment, October 1933

2/Unemployment, October 1938

Hitler visits the Berlin Motor Show in 1934. A keen supporter of motor racing, Hitler put the motor industry at the centre of the economic revival.

fit directly. To avoid any inflationary threat and to encourage high profits and high investment the government controlled wages and prices. Living standards for most Germans failed to recover the levels of 1929.

The changing role of the state after 1933 did not create a fully planned economy like Stalin's USSR, but it produced a system which German economists called a 'managed economy' (gelenkte Wirtschaft), with high levels of state regulation and intervention. Economic revival was real enough but at the cost of the more liberal economy which had produced the slump.

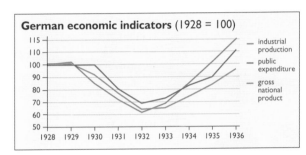

German economic indicators (1928 = 100)

— industrial production
— public expenditure
— gross national product

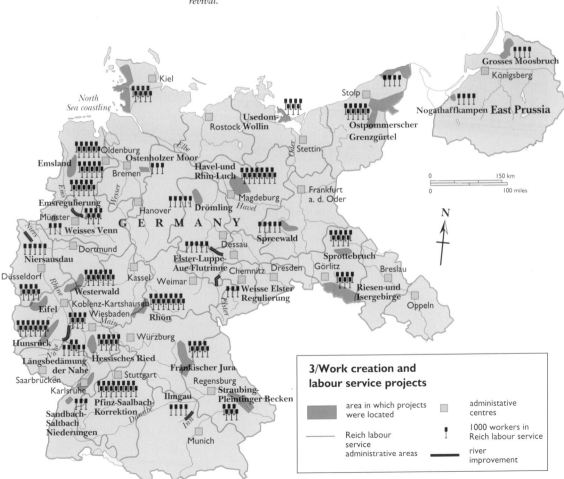

3/Work creation and labour service projects

area in which projects were located

—— Reich labour service administrative areas

administative centres

▮ 1000 workers in Reich labour service

river improvement

The Structure of Repression

The Nazi regime began to construct a system of repression and political surveillance within weeks of taking power. By 1936 the network of police and SS terror covered the whole Reich.

"You can't make an omelette without breaking eggs!"
Hermann Göring, 1933

Heinrich Himmler (1900-1945) head of the SS from 1929, became the chief of all German police and security services in 1936.

The Nazi regime imposed two forms of repression on its political opponents and other dissidents. The first developed through state channels; the second came from the activities of Party institutions, primarily Himmler's *Schutzstaffeln* (SS), but also the Party security organization, the *Sicherheitsdienst* (SD).

The framework for state repression was supplied by the Emergency Decree that followed the Reichstag Fire in February 1933. A state of emergency remained permanently in force. It allowed the police to take political suspects into 'protective custody'. Göring initiated a new secret police force to root our political resistance in the same month, which was formally constituted as the Gestapa (Secret Police Office) on 26 April 1933. Similar forces were set up in other states and gradually centralized under the control of Himmler and his deputy Reinhard Heydrich. On 17 June 1936 Himmler was appointed Chief of German Police, with extraordinary powers over the whole population. In September 1939 the system was further rationalized when the *Reichssicherheitshauptamt* (RSHA) was established to draw together the whole security apparatus of state and Party into one umbrella organization.

The repression of political enemies of the new regime produced a large prison population which was

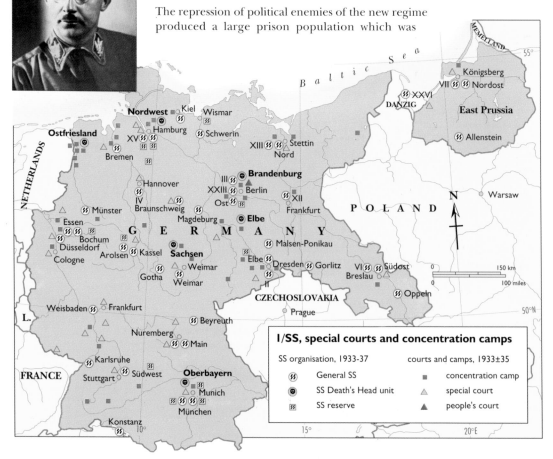

I/SS, special courts and concentration camps

SS organisation, 1933-37

Ⓢ	General SS
◉	SS Death's Head unit
Ⓢ	SS reserve

courts and camps, 1933±35

■	concentration camp
△	special court
▲	people's court

The Work of the Gestapo

**Düsseldorf Gestapo, 1933–1945:
cases by category**

	Percentage of total
Continuation of outlawed organizations (political parties, religious sects, youth groups):	30.0 %
Non-conforming behaviour (verbal utterances, work or leisure activity):	29.0 %
Acquiring or spreading forbidden printed matter:	4.5 %
Listening to foreign radio:	2.3 %
Political passivity:	0.9 %
Conventional criminality:	12.0 %
Others (unspecified):	21.3 %

**Würzburg Gestapo, 1933–1945:
cases of race crimes**

Reports from the general population:	57.0 %
Information from control organizations:	5.0 %
Observation by Gestapo informers:	0.5 %
Information from statutory authorities:	0 %
Statements extracted by interrogation:	15 %
Information from NSDAP organizations:	9.0 %
Others (unspecified):	13.5%

housed in a number of concentration camps set up from 1933. Between 1933 and 1939 approximately 225,000 Germans were imprisoned for political crimes, as defined by the regime. In 1939 there were an additional 162,000 in 'protective custody' without trial. Sentences varied from a few weeks to life. A great many of those seized by the Gestapo were the victim of denunciation by colleagues, neighbours or family. Some of these were simply malicious. The effect was to create a chaotic jumble of cases which the understaffed Gestapo offices could barely cope with. At its peak the whole security apparatus comprised 50,648 personnel for a population of over 90 million. The structure of repression relied on a great deal of 'self-policing' by society as pressure was imposed on individuals to conform. Political resistance survived in urban areas with a strong socialist tradition, but it was pushed far underground as the risks of betrayal and police terror became more marked. The social-democrats set up an exile organization based at first on Czechoslovakia, but later moved to London. Their political reports showed an initial desire by workers not to conform, followed by a long period of demoralization and political passivity. Dissent ceased to be a collective activity, but relied instead on the conscience and energy of individuals.

An example of unofficial paper currency made and circulated in the concentration camps.

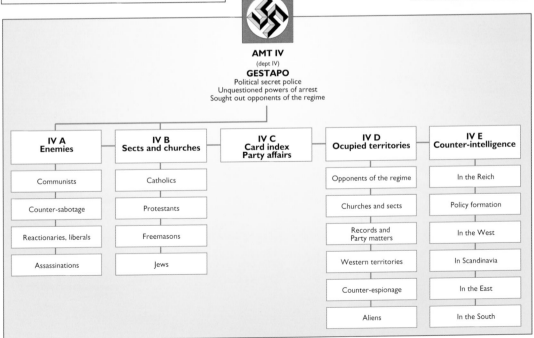

Race and Racism in the 1930s

The Nazi regime pursued a programme of 'biological politics' to create a healthy German race and to stamp out alien elements.

"Mental patients are a burden upon Germany and we only want healthy people."
Christian Wirth, commandant of Belzec camp

A poster for the anti-semitic film and exhibition 'The Eternal Jew', 1937.

The Nazi movement viewed the new Germany predominantly in racial terms. Race policy had both positive and negative aspects. It was concerned to build a healthy Germanic people, destined to be one of Nature's higher species. But it was also the task of the regime to root out those biologically undesirable elements within the German race, to prevent biological degeneration.

This concept of biological purity had its roots in the theories of racial hygiene (eugenics) popular with sections of the medical establishment in Europe and America. Eugenic theory suggested that human populations, like those in the animal kingdom, were subject to the laws of natural selection outlined by Darwin. A healthy race required the elimination of those who had physical or mental defects, or who introduced alien blood into the traditional racial stock. This pseudo-scientific view of racial policy was expressed by Hitler in *Mein Kampf*. Once in power he established an apparatus of laws and offices whose task was to cleanse the race.

On 26 July 1933 a Law for the Prevention of Hereditarily Diseased Progeny was announced, which allowed the state compulsorily to sterilize anyone deemed to be a threat to the biological health of the population. In 1936 a Reich Committee for Hereditary Health Questions was established to oversee the eugenic programme, and in the summer of 1939 Hitler formally approved the so-called 'euthanasia' action, the killing of the physically and mentally handicapped. The state also proceeded against prostitutes, abortionists and homosexuals for crimes against the race. In 1936 a Reich Office for the Combating of Homosexuality and Abortion was set up. Altogether approximately 50,000 homosexuals were punished under Section 175 of the Penal Code, and 5,000 were sent to concentration camps.

Anti-semitism also intensified. Many Jews were hounded from office or imprisoned in the first wave of lawless anti-semitism in 1933. In September 1935 at the Party Congress the anti-semitic Nuremberg Laws were

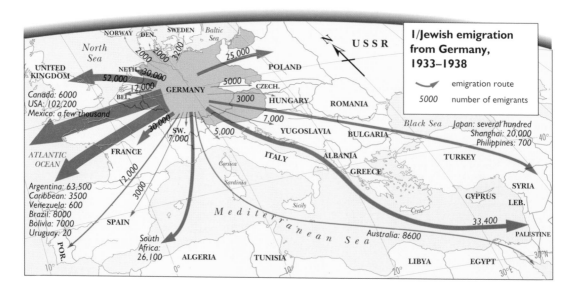

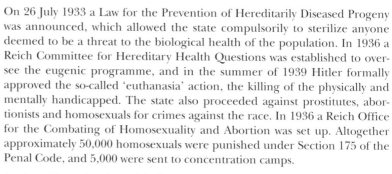

I/Jewish emigration from Germany, 1933–1938

emigration route

5000   number of emigrants

North Sea
NORWAY   DEN.   SWEDEN   *Baltic Sea*
2,000   2,000   3,200
25,000
USSR
UNITED KINGDOM
NETH.   30,000
POLAND
52,000
Canada: 6000
USA: 102,200
Mexico: a few thousand
12,000   GERMANY
BEL.   5000   CZECH.
3000   HUNGARY   ROMANIA
7,000
Black Sea   Japan: several hundred
Shanghai: 20,000
Philippines: 700
40°
ATLANTIC OCEAN
FRANCE
30,000
SW.   5,000   YUGOSLAVIA   BULGARIA
7,000
Corsica   ITALY   ALBANIA
GREECE
TURKEY
Sardinia
Argentina: 63,500
Caribbean: 3500
Venezuela: 600
Brazil: 8000
Bolivia: 7000
Uruguay: 20
12,000
3000
SPAIN
Sicily
Crete
SYRIA
CYPRUS
LEB.
Mediterranean Sea
33,400
South Africa: 26,100
POR.   ALGERIA   TUNISIA
Australia: 8600
LIBYA   EGYPT
PALESTINE
30°N
10°   0°   10°   20°   30°E

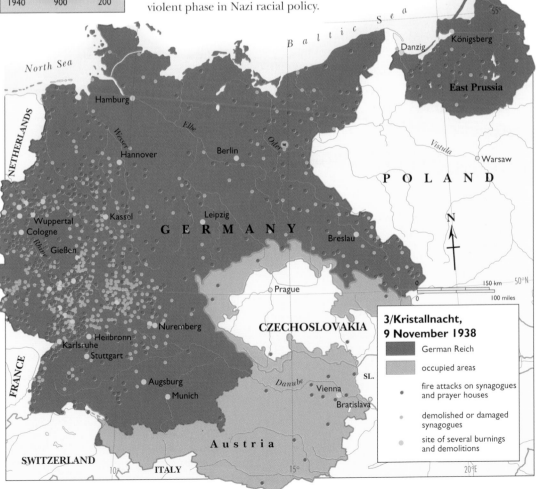

• Zionist retraining centres,
August 1936

2/Jewish emigration to Palestine		
Year	from Germany	from Austria
1933	7,600	-
1934	9,800	-
1935	8,600	-
1936	8,700	-
1937	3,700	-
1938	4,800	2,200
1939	8,500	1,700
1940	900	200

announced. The subsequent Reich Citizenship Law of 14 November 1935 defined who was and was not a Jew. The Law for the Protection of German Blood and Honour published the same day forbad intermarriage and sexual relations between Jews and Germans, but also covered relations with blacks, and the Sinti and Roma (gypsies). These laws linked the eugenic programme with the regime's anti-semitism. Over the next four years the Jewish community was gradually excluded from business and the professions (through a programme known as 'aryanization'), lost citizen status and entitlement to a number of welfare provisions. The aim of the regime was to encourage Jewish emigration. Approximately half of Germany's Jews did emigrate between 1933 and 1939, 41,000 of them to Palestine under the terms of the Ha'avarah Agreement made with Zionist organizations in Palestine on the transfer of emigrants and their property from Germany. In an unlikely collaboration with the SS, training camps were set up in Germany for emigrants to acquire the skills needed in their new life in Palestine. This process slowed down by the late 1930s as the receiver states limited further Jewish immigration. At the same time anti-semitic activity in Germany intensified. On 9 November at the instigation of leading racists in the movement a nationwide pogrom destroyed thousands of synagogues and Jewish businesses. In all 177 synagogues were destroyed and 7,500 shops. *Kristallnacht*—the 'Night of Broken Glass'—signalled the start of a more violent phase in Nazi racial policy.

3/Kristallnacht, 9 November 1938

- German Reich
- occupied areas
- • fire attacks on synagogues and prayer houses
- • demolished or damaged synagogues
- • site of several burnings and demolitions

Industry and Labour in the Third Reich

During the pre-war period both industry and labour were brought closely under state supervision and control, in the name of national unity.

"Soon Germany will not be any different from Bolshevik Russia" Fritz Thyssen, steel industrialist, 1940

One of the professed aims of Nazi ideology was to eliminate selfish class interests, characteristic of capitalist society, and to replace them with corporate organizations which reconciled the once conflicting elements of the German population. This was the thinking behind the establishment of the German Labour Front and the Reich Economic Chamber.

The Labour Front was formally established on 10 May 1933, following the seizure of trade union offices and assets on 2 May. It was run by a Party hack, Robert Ley, and it was intended to include all productive workers, including white-collar and managerial staff. Strikes and lockouts were prohibited, and labour relations were regulated by Reich Trustees of Labour, appointed on 19 May 1933 to oversee and enforce conditions of employment and labour contracts. At factory level the Labour Front in co-operation with the Nazi factory cell organization (NSBO) organized Trustee Councils elected from among the workforce from a Nazi-approved list. The existing structure of labour relations was modified by the Law on the Organization of National Labour, promulgated on 16 January 1934. Wages were fixed at the rates they had reached at the trough of the depression, and wage control remained a permanent feature of the regime's labour policy through to 1945.

The abolition of trade unions and collective wage bargaining left labour with little opportunity to promote its interests. Living standards slowly revived as

A propaganda poster showing working-class support for the Nazi leader.

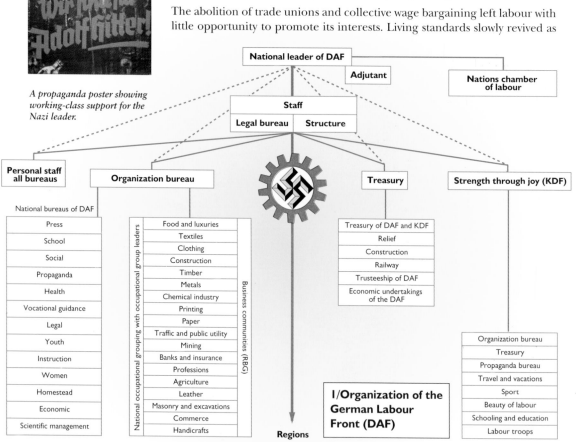

National leader of DAF

Adjutant

Nations chamber of labour

Staff

Legal bureau | **Structure**

Personal staff all bureaus

Organization bureau

Treasury

Strength through joy (KDF)

National bureaus of DAF

| Press |
| School |
| Social |
| Propaganda |
| Health |
| Vocational guidance |
| Legal |
| Youth |
| Instruction |
| Women |
| Homestead |
| Economic |
| Scientific management |

National occupational grouping with occupational group leaders

| Food and luxuries |
| Textiles |
| Clothing |
| Construction |
| Timber |
| Metals |
| Chemical industry |
| Printing |
| Paper |
| Traffic and public utility |
| Mining |
| Banks and insurance |
| Professions |
| Agriculture |
| Leather |
| Masonry and excavations |
| Commerce |
| Handicrafts |

Business communities (RBG)

| Treasury of DAF and KDF |
| Relief |
| Construction |
| Railway |
| Trusteeship of DAF |
| Economic undertakings of the DAF |

| Organization bureau |
| Treasury |
| Propaganda bureau |
| Travel and vacations |
| Sport |
| Beauty of labour |
| Schooling and education |
| Labour troops |

Regions

1/Organization of the German Labour Front (DAF)

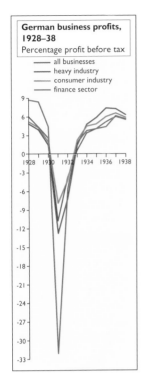

German business profits, 1928–38

Percentage profit before tax

— all businesses
— heavy industry
— consumer industry
— finance sector

more workers were brought back into employment and hours of work expanded. By 1938 real earnings were, on paper, back to the levels of the late 1920s, but compulsory Party dues and welfare contributions, and the declining quality of many food and consumer products, reduced the apparent growth of living standards. Workers in heavy industry and armaments were rewarded with bonuses or social payments, and highly skilled workers could still put veiled pressure on employers, but for workers in other less favoured sectors the economic recovery offered stable employment rather than higher incomes.

The situation for businessmen was very different. After years of crisis the Nazi economic recovery rescued many from the edge of bankruptcy. Profits grew rapidly in the areas fovoured by state contracts, though dividends were limited to 6% to encourage high rates of re-investment. The price industry paid was the loss of economic independence. Like labour, business was organized in new corporate organizations. The Reich Economic Chamber covered the whole productive economy. Its powers were never clearly defined, but membership was compulsory. Each industrial and trade branch was also organized in compulsory economic groups, centred on the national Reich Group Industry. By a law in 1933 cartels became compulsory, and prices were subject to review by a state Price Commissioner. Over 1,600 new cartel agreements were made, and by 1936 two-thirds of German industry was cartellized. The state supervised and regulated overseas trade, capital investment, the supply of scarce materials, and price and wage policy. Many businessmen resented the loss of economic independence. Fritz Thyssen, the iron baron who welcomed the advent of Hitler in 1933, fled to Switzerland in 1939 in protest at the growth of a state-led economy. Businessmen, like workers, had to find ways of circumventing the web of controls to their advantage.

2/Organization of the Reich economic chamber

▲ regional group

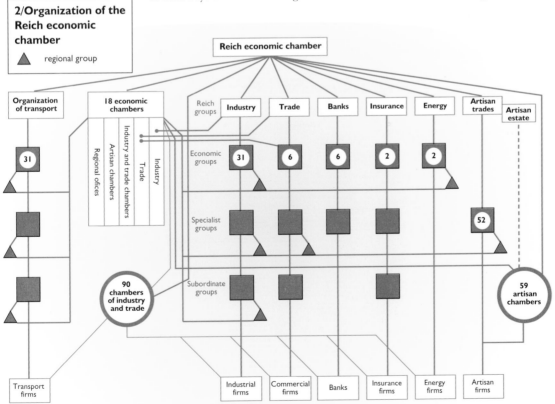

Farming in the Third Reich

The Nazi movement was ideologically committed to preserving the German peasantry, but food production was also vital to the waging of war.

"The future of the nation is solely dependent on the maintenance of the farmer."
Adolf Hitler,
October, 1933

No social group gave greater support to the Nazi Party before 1933 than the German peasantry. They were won over by promises of tariff protection and lower taxes, and by the propaganda of the Nazi racists and agrarian romantics who saw the rural population as the foundation stone of a pure and healthy national stock.

Some at least of the pre-1933 promises were redeemed. Between 1934 and 1938 farmers got 60 million marks of tax relief and 280 million marks through lower interest rates. Prices, which had fallen disastrously during the slump, were stabilized and foreign products discriminated against. Gross farm income increased from 4.2 billion marks in 1933 to 5.7 billion in 1937. These improvements prevented the decline of German farming from going any further, but it did not produce an economic miracle. Farm income grew more slowly than industrial incomes and profits. Indebtedness, if cheaper to maintain, nonetheless continued to rise during the 1930s. Labour left the land, attracted by better wages in the city, and had to be supplemented during the harvest season by gangs of Hitler Youth or men on Labour Service.

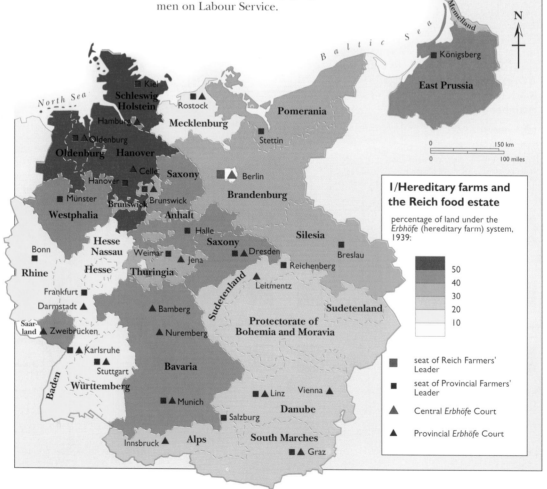

1/Hereditary farms and the Reich food estate

percentage of land under the *Erbhöfe* (hereditary farm) system, 1939:

	50
	40
	30
	20
	10

■ seat of Reich Farmers' Leader

■ seat of Provincial Farmers' Leader

▲ Central *Erbhöfe* Court

▲ Provincial *Erbhöfe* Court

Percentage of major foodstuffs produced within Germany	1927–28	1933–34	1938–39
bread grains	79	99	115
potatoes	96	100	100
vegetables	84	90	91
sugar	100	99	101
meat	91	98	97
eggs	64	80	88
fats	44	53	57
all food	68	80	83

Peasant girls in folk costume greet the Führer at the Nuremberg Party rally in 1935. The Party won over the farmers with their 'Blood and Soil' propaganda.

Farming, like other economic sectors, was subject to close regulation. Market and price controls were introduced and the state encouraged the rationalization and professionalization of farm business, which many peasants resented. The legislation that caused the most concern was the Reichserbhofgesetz (Reich Hereditary Farm Law) published on 15 May 1933. The law affected all farms of 7.5–125 hectares. Applied first to Prussia it was extended to the whole Reich in September 1933. Its purpose was to preserve the traditional peasant farm, which could not be sold or otherwise alienated, and which had to be passed on to just one named heir. In effect the law tied peasants to the farm. Special Erbhof courts were set up to decide in cases of dispute or to remove farms from unsatisfactory owners. About 35% of the farms in Germany were classified as Erbhöfe, and their owners were allowed to use the traditional word *Bauer*, which now applied only to them.

The effects of the new law were to discourage a more thorough reform of farm methods. German agriculture remained predominantly small-scale and labour intensive. Output grew only slowly, though less so in the late 1930s as the Four Year Plan worked for greater self-sufficiency in preparation for war. By 1939, with the exception of fats, Germany was close to a position of effective self-sufficiency in food. During the war home output declined, and to maintain the ration in Germany meat, fats and grains had to be imported from occupied and allied Europe. No serious food shortage developed, partly as a result of the much closer supervision exercised over agriculture than in the war of 1914–18.

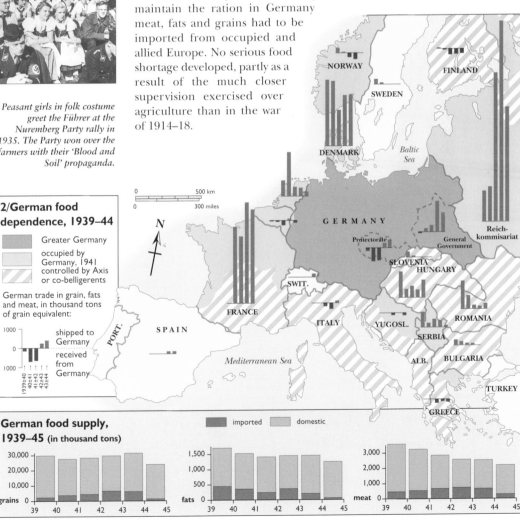

2/German food dependence, 1939–44

- Greater Germany
- occupied by Germany, 1941
- controlled by Axis or co-belligerents

German trade in grain, fats and meat, in thousand tons of grain equivalent:

shipped to Germany

received from Germany

1939±40 40±1 41±2 42±3 43±4

German food supply, 1939–45 (in thousand tons)

imported   domestic

grains: 39 40 41 42 43 44 45

fats: 39 40 41 42 43 44 45

meat: 39 40 41 42 43 44 45

Culture and Education

The Nazi regime placed great emphasis on education and popular culture as a means developing racial awareness.

"In my new training schools young people will grow who will shock the world."
Adolf Hitler

During the period of the Third Reich education and culture were both used by the regime for political purposes, as vehicles for propaganda and for the physical and intellectual development of the race. In both schools and universities there was a large proportion of Party members among the teaching staff. At school lessons on race and German history reflected the ideological imperatives of the regime. At the highest level the number of students at university and at technical college declined, partly as a result of pressure to exclude women from higher education. Enrolment at universities fell by 57% between 1933 and 1939. Most young German men on finishing school were enrolled in the compulsory Labour Service and for military training. At school level the system changed very little, but the Party introduced a separate group of élite schools where future leaders of the race and movement were given a special physical and political training. The Adolf Hitler Schools set up in 1936 were to produce the Party's future élite. There were 600 pupils in all; only 28% of them in 1940 came from big city areas, and 20% came from the countryside. The 21 national Political and Training Institutions had a strong Party bias, but were designed to produce the country's leadership corps for military, administration and party. The SA established a special Reich high school for the exceptional offspring of Party loyalists.

Throughout the schools' activities great emphasis was placed on physical education. Healthy, active bodies were regarded as neccessary for the biological welfare of the race and sporting achieve-

German neo-classical sculpture at the new Olympic Stadium in 1936. The regime encouraged heroic statuary and classical images and outlawed 'degenerate' modern art.

1/Adolf Hitler schools and Hitler Youth

- ▨ Greater German Reich
- ● national Adolf Hitler school
- ■ national educational institute
- Ⓢ Reich school
- ⬛ Reich school of the NSDAP

2/SA sport badges awarded, 1933–37

Per 1000 inhabitants:

	20
	18
	16
	14

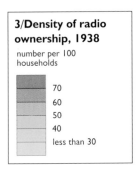

A poster issued by the Propaganda Ministry encouraging radio ownership.

3/Density of radio ownership, 1938

number per 100 households

	70
	60
	50
	40
	less than 30

ment was heralded as a racial duty. Thousands of young Germans took the Reich sports tests to qualify for the Reich Sports Badge. Programmes of gymnastics and callisthenics were introduced in schools, offices and factories. The Strength through Joy organization of the Labour Front employed 1,000 full-time sports instructors. When Germany bid successfully to host the Olympic Games in 1936 the regime was determined to use the occasion as a propaganda triumph for the regime. Despite threats of boycott because of German ant-semitism in sport (which was introduced as early as April 1933 when Jews were banned from Germany's 13,000 gymnastic clubs), the games went ahead. German sportsmen dominated, winning 33 golds to the USA's 24, though to Hitler's dissappointment they did not win a single track event.

As part of the preparation for the games, Goebbel's Propaganda Ministry, set up in 1933, undertook to establish a worldwide broadcasting network to relay the Games as they happened. Control of broadcasting was a major feature of the propaganda effort of the regime. In 1933 alone 50 Hitler speeches were relayed over the radio. In May 1933 work began on developing the mass-produced 'People's Radio' (*Volksempfänger*) of which there were 3.5 million sets by 1939. By then 70% of German households possessed a radio, while the Goebbels ministry planned to set up 6,000 loudspeaker towers in city streets to bring propaganda directly to the people. Radio was also used as a foreign policy tool. German stations bombarded the German-speaking populations of Austria, Poland and Czechoslovakia with nationalist broadcasts as part of a campaign of destabilization. Goebbels succeeded in putting propaganda at the centre of German political and cultural life.

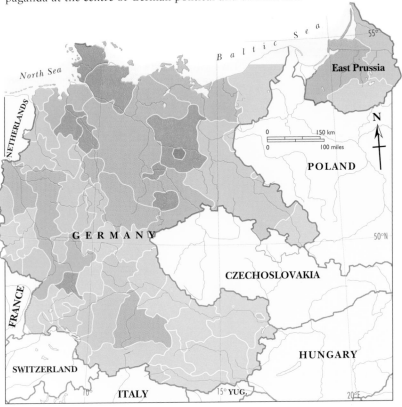

Constructing the New Germany

Hitler hoped to use architecture as a means to ensure the survival of the new Reich and its values and as a symbol of its power.

"The whole world will come to Berlin to see our buildings"
Adolf Hitler, 1938

Hitler accompanied by the Party's pet architect, Albert Speer, (right) on an inspection of the building of the House of German Art in Munich, 1934.

Architecture remained Hitler's greatest pleasures in the years of power. Even in the last days in the bunker in Berlin he kept plans and models of the rebuilding of Germany's cities to distract him from the grim reality outside. He saw himself as a literal builder of the new Germany in 1933. His model was ancient Rome, whose strength was built in his view on its roads and the splendid public buildings of the Empire.

Once in power, Hitler searched for architects who would share his obsession with the physical reconstruction of the Reich. The first work was led by Paul Von Troost, but in 1936 Hitler asked the young architect Albert Speer, who had designed the new Reich Chancellory in 1934, to take over the rebuilding of Berlin. By 1939 Speer was the Reich's architectural leader. In addition to rebuilding Berlin, Speer was to complete the Party City of Nuremberg—where the giant stadium for party rallies was to accommodate 250,000 people—and to build a permanent home for the Olympic Games, which Hitler expected after the war to host in perpetuity. In Austria his hometown of Linz was to become a vast new capital, combining a new industrial centre with major civic buildings. Everything was conceived on a colossal scale. Hitler assumed that the vast neo-classical buildings would instil in the beholder a sense of awe at the power that built them.

The new cities were to be linked by multi-lane motorways, the Reichsautobahnen. Building on a number of pre-1933 experiments, Hitler ordered a national network to be started in 1934 under the direction of the Nazi engineer, Fritz Todt. The planned network of over 14,000 kilometers was, on Hitler's insistence, to be carefully landscaped as a permanent monument to the aesthetic concerns of the Führer. By 1939 only 3,077 kilometers were completed, much of it on the so-called Party route from Berlin to

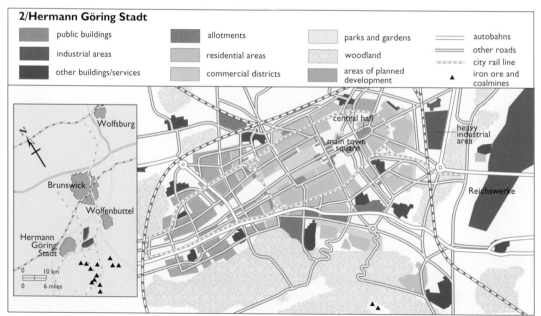

2/Hermann Göring Stadt

- public buildings
- industrial areas
- other buildings/services
- allotments
- residential areas
- commercial districts
- parks and gardens
- woodland
- areas of planned development
- autobahns
- other roads
- city rail line
- ▲ iron ore and coalmines

Wolfsburg

central hall

heavy industrial area

main town square

Brunswick

Wolfenbuttel

Reichswerke

Hermann Göring Stadt

0   10 km

0   6 miles

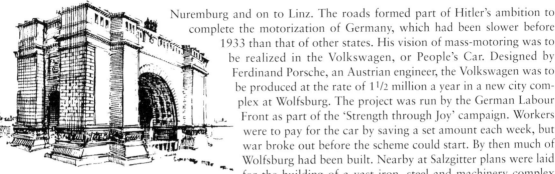

Nuremburg and on to Linz. The roads formed part of Hitler's ambition to complete the motorization of Germany, which had been slower before 1933 than that of other states. His vision of mass-motoring was to be realized in the Volkswagen, or People's Car. Designed by Ferdinand Porsche, an Austrian engineer, the Volkswagen was to be produced at the rate of 1 1/2 million a year in a new city complex at Wolfsburg. The project was run by the German Labour Front as part of the 'Strength through Joy' campaign. Workers were to pay for the car by saving a set amount each week, but war broke out before the scheme could start. By then much of Wolfsburg had been built. Nearby at Salzgitter plans were laid for the building of a vast iron, steel and machinery complex which was to form the centre of Herman-Göring-Stadt, another city (below left) of the economic boom. Wolfsburg and Salzgitter were the vanguard of a wider programme to shift the main weight of the industrial economy away from the Ruhr to central and eastern Germany. Here it was planned to set up the industrial heartland of the New Order in Europe.

Above: Hitler's sketch from 1925 of a gigantic triumphal arch, used in later plans for the remodelling of Berlin.

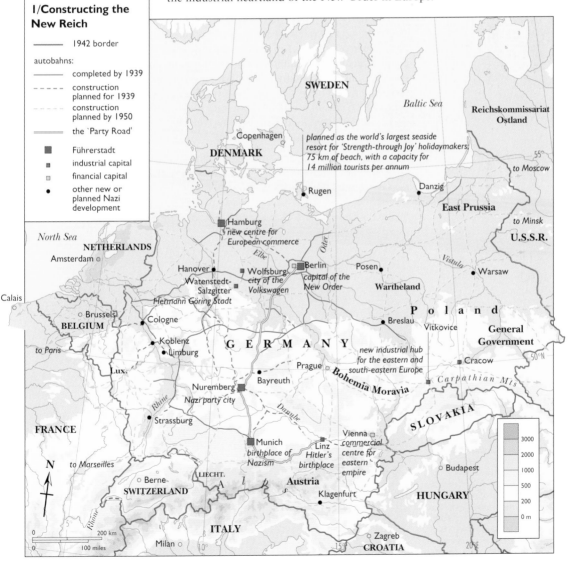

1/Constructing the New Reich

——— 1942 border

autobahns:

——— completed by 1939

– – – construction planned for 1939

- - - construction planned by 1950

——— the 'Party Road'

■ Führerstadt

■ industrial capital

□ financial capital

● other new or planned Nazi development

planned as the world's largest seaside resort for 'Strength-through-Joy' holidaymakers; 75 km of beach, with a capacity for 14 million tourists per annum

SWEDEN

Baltic Sea

Reichskommissariat Ostland

Copenhagen

DENMARK

55°

to Moscow

Rugen

Danzig

East Prussia

to Minsk

U.S.S.R.

North Sea

NETHERLANDS

Amsterdam

Hamburg
new centre for European commerce

Hanover

Wolfsburg *city of the Volkswagen*

Watenstedt-Salzgitter
Hermann Göring Stadt

Berlin *capital of the New Order*

Posen

Warsaw

Wartheland

Calais

BELGIUM

Brussels

Cologne

Koblenz

Limburg

GERMANY

Breslau

Vitkovice

P o l a n d

General Government

to Paris

Lux.

Prague

new industrial hub for the eastern and south-eastern Europe

Cracow

50°N

Nuremberg
Nazi party city

Bayreuth

Bohemia Moravia

Carpathian Mts

FRANCE

Strassburg

SLOVAKIA

N

to Marseilles

Munich
birthplace of Nazism

Linz
Hitler's birthplace

Vienna
commercial centre for eastern empire

Budapest

Rhine

LIECHT.

A l p s

Austria

HUNGARY

Berne

SWITZERLAND

Klagenfurt

3000
2000
1000
500
200
0 m

0 ____ 200 km
0 ____ 100 miles

ITALY

Milan

Zagreb

CROATIA

20°E

III: Foreign Policy in Germany, 1933–193

Between 1933 and 1939 Germany embarked on a massive programme of remilitarisation and an active foreign policy.

"Every healthy, vigorous people sees nothing sinful in territorial acquisition, but something quite in keeping with nature…"
Adolf Hitler, 1928

When Hitler came to power in 1933 there was no sharp change in German foreign policy, despite the noisy revisionist rhetoric of Party activists. The Foreign Minister remained the same, a career diplomat appointed in June 1932, Freiherr Constantin von Neurath. He was, like much of the German foreign service, a firm nationalist. Even before Hitler's appointment Germany had begun to free herself from the constraining bonds of the Versailles settlement. Under his stewardship German foreign policy continued to be directed at recovering German sovereignty and restoring German parity with the other great powers.

The first sign of a new direction came in October 1933 when Hitler took Germany out of the League of Nations and the Disarmament Conference at Geneva on the grounds that Germany was still treated as a second-class citizen in the other capitals of Europe. Hitler agreed to plans for expanding the military already laid before 1933, and authorized the secret development of an air force. Forms of military training were introduced in youth movements and the numerous 'gliding clubs' which provided the first volunteers to fly Germany's forbidden aircraft. But Hitler approached foreign policy issues prudently. German leaders did not want to provoke Allied intervention. Like every state affected by the slump, the Hitler regime wanted to stabilize and revive economic life before pursuing a more active policy abroad.

The main strands of German foreign policy in the first three years of Hitler's chancellorship were remarkably different from the foreign policy constellation created by the late 1930s. In *Mein Kampf* and in his second book written in 1928 Hitler argued that German interests would best be served by an alliance with Britain, which would give Germany a free hand in Continental

Hitler Youth undergo small arms training under the watchful eye of an HK section leader. The regime instituted limited para-military training in all youth groups, and for young Germans in the Labour Service, who were drilled with spades instead of rifles. By the time conscription was introduced in 1935 many young Germans were already familiar with the rudiments of military life.

A contemporary painting of a German infantry soldier in full uniform, in 1936. The Nazis were able to capitalise on popular militarism in German society. Many of the propaganda images of the regime had a strongly militarist aspect. When conscription was re-introduced in 1935 the army planned to build a force of 300,000. By 1945 more than 13 million Germans had been in uniform.

Europe, particularly in the conquest of 'living-space' (*Lebensraum*) in the east. After 1933 he looked for an improvement in German-British relations. Economic agreements over trade and payments meant that Britain was almost the only major state to keep credit lines open to Germany in the 1930s. Even by 1939 Britain had investments in Germany worth over £60 million. In 1935 the two countries reached an agreement over naval armaments. Britain did little in response to Hitler's declaration in 1935 that Germany was officially rearming. By contrast relations with Italy and Japan were nothing like as favourable as they were to become. The German Foreign Office preferred links with China, where Germany had large trading interests. Italy looked at the resurgent Germany to the north with some anxiety, and only the rift with the western European powers over the Ethiopian war of 1935/6 drove Mussolini towards closer ties with Germany. In the east Germany concluded a Non-Aggression Pact with Poland in 1934, but as an enemy of Marxism enjoyed only distant and hostile relations with the USSR, the very reverse of the situation five years later.

The re-orientation of German foreign policy co-incided with the decision taken in 1936 to speed up the pace of rearmament and to embark on a programme of large scale import substitution, or autarky. In the summer of 1936 following the successful re-occupation of the demilitarized zone in the Rhineland in March, Hitler drafted the so called Memorandum on the Four Year Plan which was announced amidst much noisy propaganda at the Nuremberg Party rally in September. Its unpublished purpose was to prepare the economy and armed forces for war by the year 1940. At about this time Hitler began to drop clear hints to those of his entourage who were privy to the thoughts of the Führer about the prospect of a great war in the 1940s which would redraw the map of Europe in Germany's favour. There was never a clearly articulated plan. Hitler was prepared to be flexible, to await opportunities. On a number of occasions he indicated that the years 1943–5 were the point at which war was likely, even desirable. This was the date he gave to his service chiefs and close colleagues at a meeting on 5 November 1937 which he called in order to give them an idea about his future strategy. The details, set down in the so-called 'Hossbach Memorandum' (a record of the meeting later compiled by Hitler's army adjutant) indicated the immediate aims of incorporating Austria and destroying Czechoslovakia.

The shift to large-scale armaments and a strategy of opportunistic expansion altered Germany's position in Europe. The other driving force behind that change was the Party's self-proclaimed foreign policy expert, Joachim von Ribbentrop. He had helped to secure the Anglo-German Naval Agreement in 1935, and was sent as Hitler's special representative to London to try to

A poster of 1936 declares 'Germany, your Colonies'. After 1933 it was expected that Hitler would demand the return of Germany's African and Polynesian territories, but he was not particularly interested in overseas possessions. During the invasion of the USSR he said 'This is our India'. Empire for Hitler meant a Eurasian 'living space'.

secure a wider British alliance. His aims, first as special commissioner, the from August 1936 as German ambassador, were frustrated by his own diplomatic ineptitude, but more seriously by the unwillingness of the British government to make any substantial concession to the German position. Ribbentrop became a strident anglo-phobe during his years in London and he succeeded in turning Hitler too against a British alliance. Ribbentrop was also keen to support Japan rather then China, and was instrumental in setting up the Anti-Comintern Pact directed against international Communism, signed on 25 November 1936. Italy joined a year later. Despite persistent distrust of German motives, Mussolini was wooed by Ribbentrop and by Herman Göring. Italy's isolation following the Ethiopian war, and tension with the western states from the summer of 1936 over Italian intervention in the Spanish Civil War, pushed Mussolini towards Germany.

No clear military or political agreement existed between Germany, Italy and Japan, but they became identified in the mid-1930s as revisionist powers, who hoped to alter the existing distribution of territory and international power in their favour. The direct result was to create a widening rift with Britain and France, which Hitler had not anticipated and did not welcome. In the spring of 1938 political crisis within Austria provided the opportunity to bring about a forced union of the two states. In May 1938 persuaded by Ribbentrop that Britain and France were decadent and isolationist, he planned to wage a swift war against Czechoslovakia in the autumn to bring the Sudeten Germans under German rule, and to seize valuable economic resources for the large German military build-up. As tension between Berlin and Prague rose during the summer months Britain and France sought to interfere in the conflict.

The Czech crisis began the long descent to war in September 1939. The British and French were not prepared to allow Hitler a free hand in eastern Europe to build a new German empire. They were willing to make what they regarded as reasonable concessions arrived at by negotiation. In September 1938 the British Prime Minister, Neville Chamberlain, flew to meet Hitler in Germany in order to mediate the transfer of German-speaking areas in Czechoslovakia to German rule. Hitler reluctantly complied, following an

Sudeten Germans organize in a 'Free Corps' in 1938 as part of the struggle to win independence for the German-speaking parts of Czechoslovakia. The Sudeten German leader, Konrad Henlein, had close links with the Nazi regime. When the area was taken over in October 1938 by the German state it was incorporated completely. Henlein became Gauleiter of Gau Sudetenland.

A German poster of 1939 charting the development of Grossdeutschland — Greater Germany. A Pan-German state had been the aim of many German nationalists since the period of unification in the mid-19th century. Bismarck, a Prussian, created a small Germany in 1871; Hitler, an Austrian, created a large Germany in 1939. Bismarck's creature lasted for 47 years, Hitler's for only 6.

appeal from Mussolini to air the issues at a Great Power conference. The meeting at Munich on 29 September 1938 gave Germany the Sudetenland, but averted Hitler's small war with the Czechs. Six months later, on the pretext that the rump Czech state was becoming ungovernable, German forces entered the country unopposed and partitioned it into a Protectorate of Bohemia and Moravia and an autonomous Slovakia. The first Reich Protector was von Neurath, Hitler's first Foreign Minister, who in February 1938 had been replaced by the ambitious Ribbentrop.

Hitler drew a number of lessons from the Czech crisis. First, he was determined not to give way again to what he saw as largely hollow threats from the west. Second he believed that the concessions made by Chamberlain amounted to a virtual green light to further expansion in the east. He developed an obsessive belief that Britain and France lacked the willpower to obstruct him further, though they would make much noisy protest. His conception of his immediate foreign policy aims became clearer. He was building the political and economic foundation in central Europe of a German super-power whose economic size and military strength would of itself produce a revolution in the European balance of-power. The next target of German pressure, Poland, was to become part of that new empire either voluntarily, by becoming a dependent state like Slovakia and, increasingly, Hungary, or involuntarily through war. When Polish leaders rejected the first course, Hitler directed the armed forces to prepare for another autumn campaign and a short sharp war.

From the point of view of European stability Poland was the most explosive choice Hitler could have made for the next stage of expansion. For in March 1939 Britain singled out Poland for a unilateral guarantee of sovereignty, which France backed up with some reluctance. Britain made it clear that if Germany violated that sovereignty then war would result. Hitler did not believe it. He asserted to his generals and to his major party leaders the view that Britain and France were bluffing, a view from which he wavered little right up to the day, 1 September, when German forces attacked Poland. But he was not a complete gambler. In early 1939 he began to explore the possibility of mending fences with the USSR, the sworn ideological enemy, as a means to avoid any risk of a two front war, or of a renewed 'encirclement' like that of 1914. In August Stalin finally took the bait after he was convinced that Britain and France had little to offer him for Soviet co-operation. On 21 August Ribbentrop flew to Moscow where he negotiated a comprehensive political and trade agreement in a matter of hours. The Nazi-Soviet Non-Aggression Pact was signed on 23 August.

Hitler used the Pact as a way of bolstering his conviction about western bluff. With no prospect of a Soviet alliance and no effective way of helping Poland militarily, Hitler could see no reason for western intervention. Others around him including Göring and Goebbels, tried to persuade him that the risk was a very great one. Ribbentrop, whose own reputation with Hitler rested on his reading of British weakness, spurred Hitler on. On 1 September invasion of Poland began. Two days later Britain and France declared war to Hitler's evident consternation. Hitler had achieved in 1939 the precise reverse of his goals in 1933: a treaty of friendship and co-operation with the USSR and a war with Britain.

Overturning Versailles, 1933–1936

Between 1933 and 1936 Hitler took the first steps to undermine the Versailles Settlement, which he was committed to overturn.

"A treaty has been ratified which promised peace but which brought in its wake endless bitterness and oppression."
Adolf Hitler, 1935

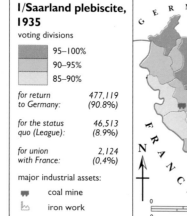

Army officers wearing gasmasks during a military demonstration on the Zeppelinfeld in 1935, the year in which German rearmament was publicly announced for the first time.

Hitler shared with most of his adopted compatriots a profound hatred of the Versailles peace settlement, and a deep resentment of Germany's military inferiority and economic subservience. During the first three years in power he defied and overturned many of the central features of the treaty.

That Hitler was able to do so owed a good deal to the changing international circumstances after the slump. Other states were absorbed with their own domestic difficulties. The economic crisis reduced the willingness of the western powers to collaborate, and discouraged any risks in foreign policy. Reparations had already been suspended in June 1932, well before Hitler came to power, and by agreement between the powers involved rather than through unilateral action on Germany's part. The occupation forces in the Rhineland were withdrawn in 1930, five years ahead of schedule. The willingness and ability of the victor powers to continue to enforce the Treaty were already in question. Hitler made it clear that no further reparations would be paid, and in doing so asserted Germany's economic independence.

Hitler's next priority was military independence. Here, too, the process of undermining the disarmament clauses in the Versailles Treaty had begun before Hitler came to power. The armed forces developed aircraft, submarines and tanks in veiled collaborative projects in the Soviet Union, Sweden and the Netherlands. In September 1928 the Defence Ministry approved the so-called First Armaments Programme, a four year programme to provide the supplies and munitions for a 16-division army. In 1932 the Second Armaments Programme was approved to prepare a 21-division army by 1938, while a 'Conversion Plan' produced in November 1932 provided for covert means to expand Germany's military personnel. These programmes formed the basis of Hitler's rearmament from 1933. In addition, in July 1933 the first plans were laid down for a new German airforce. A 26-squadron force was prepared in secret, camouflaged by the civil aviation facilities which temporarily housed them. The first requirement was for trainer aircraft to create the new cadres of flying personnel. In 1934 the secret Luftwaffe possessed some 1,300 trainer aircraft, but only 99 fighters and 270 bombers, many of them crudely converted civil aircraft.

Rearmament in the first two years of the Hitler regime was relatively modest in scope, though it violated the disarmament clauses of Versailles. Hitler and the armed forces proceeded cautiously, partly from fear of foreign intervention, partly from fear of the economic consequences of rapid remilitarization, but mostly

I/Saarland plebiscite, 1935

voting divisions

	95–100%
	90–95%
	85–90%

for return to Germany:	477,119	(90.8%)
for the status quo (League):	46,513	(8.9%)
for union with France:	2,124	(0.4%)

major industrial assets:

- coal mine
- iron work

(map) GERMANY · Saarland · Saarlouis · Neunkirchen · Saarbrücken · FRANCE · N

0 — 200 km
0 — 100 miles

fighter group flew from Lippstadt, landed at Köln-Butzweilerhof; one squadron flew on to Düsseldorf

2 March order given for 'Winter Exercise'

7 March 3 battalions of infantry crossed Rhine bridges

fighter group 134 flew from Kitzingen, sending 2 squadrons to Frankfurt and 1 squadron to Mannheim

7–9 March garrisons set up in Aachen, Trier, Saarbrücken

2/Rhineland remilitarization, 1936

— border of Germany

demilitarized zone

German territories occupied 1923–30

area of Saar plebiscite 1935

German advance

because they wanted a planned, step-by-step development of a military infrastructure that had been largely dismantled in 1919–20. By 1935 Hitler was prepared to go further. On 16 March he formally announced the re-establishment of Germany's armed forces and German rearmament. Plans were finalized for a 36-division, 700,000-man army and an air force of more than 200 squadrons. There was little international protest. Three months later Britain signed a Naval Agreement with Germany limiting total German tonnage to 35% of the Royal Navy, but endorsing Germany's right to rearm at sea.

The same year the territorial settlement of 1919 was altered in Germany's favour. As agreed in the treaty a plebiscite was held in the Saar region to determine the future status of the area. The plebiscite was policed by a League of Nations force of British, Dutch, Italian and Swedish troops, and following a frantic campaign of propaganda over 90% of the population voted for union with Hitler's Germany. A year later Hitler gambled that he could re-occupy the demilitarized Rhineland with German forces without a serious foreign threat. On 7 March 1936 troops crossed the Rhine bridges and Germany fully regained her military independence.

4/Luftwaffe organization to 1935

demilitarized zone

Luftwaffe regional command H.Q

IV Luftwaffe regional command

airbase built or building

aircraft manufacturing plant

engine plant

3/Army organization to 1935

army group H.Q.

army H.Q.

demilitarized zone

IV Wehrkreis (military district)

The Nazi Party Abroad

The Nazi Party used German expatriate communities in Europe and the Americas to further their political interests abroad.

"The German component of the American people will be the source of its political and mental resurrection."
Adolf Hitler, 1933

The leader of the pro-Nazi German-America League, Fritz Kuhn, addressing a meeting under the watchful eye of George Washington. He was charged with embezzling funds in 1939 and imprisoned.

In 1939 it was estimated by German authorities that between 10 and 11 million Germans—defined as those whose mother-tongue was still German—lived abroad. There were three million German-Americans, but the bulk lived in Europe and the Soviet Union. The Nazi movement saw these ex-patriate communities as potential Nazi supporters, and set up a party foreign department (*Auslands-Organization*), AO, to establish contacts and to co-ordinate the political activities of German nationalists abroad. The organization had eight departments defined by geographical area, under the control of Ernst Bohle. Its record abroad was mixed. In those areas with large German minorities brought about by the peace settlement (Czechoslovakia, Poland) the AO worked like the party to achieve a pan-German state. In more distant areas the activities of the Party were often regarded with hostility.

In the Free City of Danzig, set up by the League of Nations in 1919, the largely German population returned assemblies before 1933 that mirrored the political divisions in Weimar Germany. In 1930 the Nazi Party gained a surprising 19% of the vote, and in May 1933 became the largest party. Danzig was dominated by its own Nazi movement down to 1939, when it was incorporated into the Reich. In the Sudetenland areas of Czechoslovakia the Nazi movement was banned in 1933, but Nazi sympathizers clustered into Konrad Henlein's Sudeten German Party and kept up pressure for union with Germany. From 9,500 members in October 1933, the Party

CANADA

UNITED STATES OF AMERICA

NORTH ATLANTIC OCEAN

Mexico

Venezuela

B r a z i l

Peru

Paraguay

Chile

Argentina

Uruguay

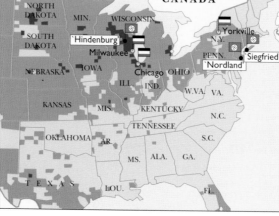

2/The Germans in the USA

percentage of Germans in the population, 1890:

| | 15 |
| | 1 |

centre of the Volksbund

training camp

NORTH DAKOTA — MIN. — WISCONSIN — CANADA
SOUTH DAKOTA — 'Hindenburg' — Milwaukee — Yorkville N.Y.
NEBRASKA — IOWA — Chicago — OHIO — Siegfried — Nordland
KANSAS — ILL. — IND. — W.VA. — VA. — PENN.
MIS. — KENTUCKY — N.C.
OKLAHOMA — AR. — TENNESSEE — S.C.
TEXAS — MS. — ALA. — GA.
LOU. — FL.

had 1.3 million members by the summer of 1938. Even in Alsace, won back by France in 1919, the Nazi party had a dozen district centres in the major towns by 1933.

With overseas communities the AO had to proceed more cautiously, since there was no question of promising a Pan-German state to Germans in the United States or Latin America. The Nazi movement in the United States (map below) was established as a party *Gau* between 1931 and 1933, recruiting mainly from the wave of emigrants to the USA in the 1920's. In 1933 Hitler switched support to the German-America League (*Amerikadeutscher Volksbund*) which was led by Fritz Julius Kuhn, a member of the German Nazi Party in 1921, who emigrated to Mexico and then the USA. The Bund had an estimated 25,000 followers, but its activities were too undiplomatic even for Hitler, who cut formal ties with Kuhn in 1935. In 1939, following violence at a Bund demonstration in New York, Kuhn was arrested on embezzlement charges and the party became subject to an investigation by the Senate Un-American Activities Committee.

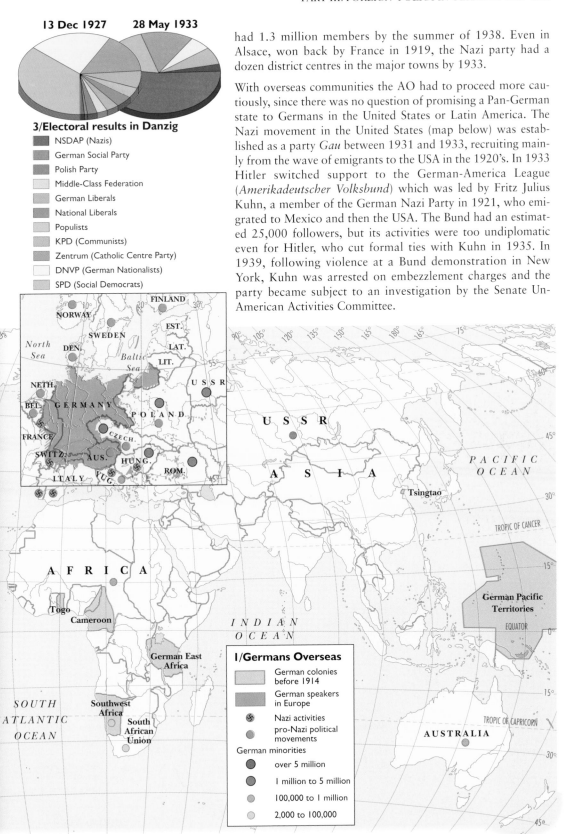

13 Dec 1927   28 May 1933

3/Electoral results in Danzig

- NSDAP (Nazis)
- German Social Party
- Polish Party
- Middle-Class Federation
- German Liberals
- National Liberals
- Populists
- KPD (Communists)
- Zentrum (Catholic Centre Party)
- DNVP (German Nationalists)
- SPD (Social Democrats)

1/Germans Overseas

- German colonies before 1914
- German speakers in Europe
- Nazi activities
- pro-Nazi political movements

German minorities

- over 5 million
- 1 million to 5 million
- 100,000 to 1 million
- 2,000 to 100,000

Bloodless Conquests 1938–1939

In 1937 Hitler turned to the question of creating a Pan-German state by absorbing the German populations of Austria and Czechoslovakia. In two years the goal was realized.

"The aim of German policy was to make secure and to preserve the racial community and to enlarge it."

Adolf Hitler,
5 November,
1937

Hitler, like all Pan-Germans, longed to unite the German peoples of central Europe into a single state. Pro-Nazi organizations existed in both Austria and Czechoslovakia among German populations anxious to share in the Hitler revolution. By 1937 Hitler felt that international circumstances and Germany's own growing military strength were sufficient to allow him to force the pace in his foreign policy. In November 1937 he warned his military leaders that a settlement of the Austrian and Czech questions was next on his agenda when the right moment came.

The opportunity to incorporate Austria into Germany arrived soon afterwards. Growing pro-Nazi agitation in Austria against the government of Kurt Schuschnigg threatened to destabilize Austrian politics. France was pre-occupied with the Spanish Civil War crisis and Italy, while Britain was still aloof from European politics; Hitler decided to act while the west was caught off-guard. In February 1938 Schuschnigg was presented with an ultimatum from Berlin to bring Nazis into his government and to align Austrian foreign and economic policy with Germany's. He refused and organized a referendum on union with Germany. Rather than risk an adverse result the Berlin government put pressure on Vienna to accept German 'protection'. On 12 March 1938 German troops

1/Austria to 1938

—— Austrian border

- - - administrative border

0 ___ 100 km
0 ___ 60 miles

Linz · Danube · Vienna
Upper Austria · Lower Austria
Salzburg
A U S T R I A · Burgenland
Innsbruck · Salzburg · Styria
Voralberg · Tyrol · Graz
East Tyrol to Tyrol · Carinthia · Mur
Villach · Jauntal
CZECHOSLOVAKIA

Franconia
Bayreuth
Württemberg · GERMANY
Swabia · Dachau
Upper Bavaria · Mauthausen · Lower Danube · Vienna
Sonthofen · Upper Danube
O S T M A R K
Tyrol - Voralberg · Salzburg · Styria
SWITZ. · Carinthia · HUNGARY
ITALY
YUGOSLAVIA

2/Ostmark, 1938

—— international border

- - - administrative border

✠ Nazi party training centre

■ concentration camp

YUGOSLAVIA

Hitler crosses the border into the Sudetenland on 3 October 1938 amidst scenes of overwhelming enthusiasm from the three million German inhabitants.

occupied Austria. Political opponents were immediately rounded up and Austria's economy and administration co-ordinated with those of Germany.

Delighted at the union—or *Anschluss*—of his homeland with his adopted country, Hitler turned to the large German minority in the Czech state left there after the breakup of the Habsburg Empire. Pleased at the lack of reaction from the other powers to the *Anschluss* he began to plan from May the military conquest of Czechoslovakia, (Case Green). The same tactics of international destabilization were practised by Henlein's Sudeten German Party. German propaganda attacks on Czechoslovakia reached a crescendo early in September a few weeks before the planned military assault. Then, on 15 September, the British Prime Minister, Neville Chamberlain, flew to see Hitler to try to avert a war. After two meetings at which Chamberlain indicated that Britain and France would not accept a German invasion, Hitler agreed to negotiation. The Sudetenland was granted to him at the Munich Conference on 28 September 1938, and three days later German troops began to occupy the German-speaking areas of Czechoslovakia. Poland and Hungary also seized the opportunity to extract territorial concessions from the Czechs. The rump Czech state was, however, in little position to resist German pressure. In March the Prague government was also forced to ask for 'protection' and on 15 March 1939 German forces occupied Bohemia and Moravia, leaving Slovakia a fragile independence as a puppet state of the Nazis.

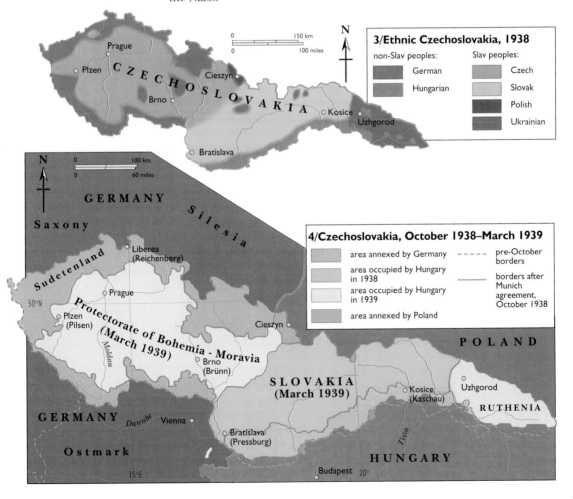

Autarky and Rearmament

In 1936 Hitler launched large scale German rearmament and the construction of a siege economy based on the policy of autarky, or self-sufficiency.

"Carrying out the armaments programme...is the task of German politics."
Hermann Göring, June 1936

Hermann Göring (1893–1946), who became Plenipotentiary for the Four Year Plan in October 1936, and was by 1938 a virtual economic dictator.

The year 1936 marked a significant break in German economic and military strategy. With the economy back to the levels of the 1920s and the political stability of the regime more assured, Hitler was in a position to start rearming seriously and to create an economy geared more closely to the conduct of large-scale warfare.

In August 1936 Hitler drafted a document on German strategy (later known as the Four Year Plan memorandum) which formed the basis for the reorientation of policy. Hitler wanted Germany to develop a 'blockade-free' economy in which materials vital for war would be produced at home instead of imported. This policy of self-sufficiency, or autarky , was introduced with the formal establishment of a Four Year Plan organization in October 1936 with Hermann Göring as its plenipotentiary. Over the following two years the organization initiated expensive investment programmes in basic chemicals, synthetic fuel oil and synthetic rubber, aluminium and iron-ore extraction, all designed to reduce German dependence on outside sources. The Plan also introduced a programme of improvement in German agriculture to ensure that Germany would not be starved into submission in a future war, and a programme of labour retraining to make sure that the depth of skilled labour needed for a war economy would be in place. Between 1936 and 1939 almost two-thirds of all industrial investment was devoted to the implementation of the autarkic plans.

Side by side with the economic strategy Hitler ordered a very great increase in military expenditure designed to bring the armed forces to a state of war-readiness by 1940. The armed forces themselves had developed plans for expansion during 1936, but Hitler's programme imposed a faster pace and a larger scale. In 1938 he ordered work to begin on a programme of explosives production which dwarfed anything produced during the First World War. Göring was told to increase armaments output by 300%; the air force 'Plan 8'

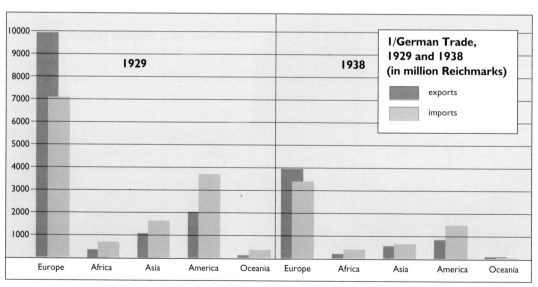

2/ Z Plan for the German Navy, 1939–47

Battleships	13
Battle cruisers	12
Aircraft carriers	8
Cruisers	65
Destroyers	70
Torpedo boats	78
Submarines	249

foresaw a five-fold increase in German air power, including a large complement of heavy bombers; army motorization was accelerated, with a target date for completion by 1942. In January 1939, following Hitler's realization that Britain and the USA might well become German adversaries over the next decade, he ordered the navy Z-Plan to begin, which would give Germany a large battle fleet by the mid-1940s.

So large were the plans for rearmament and economic development that neither the armed forces nor the economic administration thought them remotely feasible. The level of state debt increased sharply and the growth of living standards came virtually to a halt. The militarization of the economy forced a much higher level of state management and eroded the independence of private business. The price of extensive military commitment was a command economy and a distortion of Germany's pattern of consumption and international trade.

4/Göring's Industrial Empire, 1940

▨	rubber factory
◉	synthetic rubber
△	aluminium plant
▢	magnesium plant
○	synthetic oil refineries
↑	oilfields
◗	iron ore
✛	aircraft production plants
⊞	aero engine plants
Junkers	aircraft/engine manufacturers

3/Military expenditure, 1933/4–1938/9

Years	Budget Expenditure (billion RM)				*Mefo bills	Total
	Total	Army	Navy	Air	*Secret military funds used to disguise rearmament	
1933/4	0.75	–	–	–	–	0.75
1934/5	1.95	0.81	0.50	0.64	2.14	4.09
1935/6	2.77	1.04	0.7	1.03	2.72	5.49
1936/7	5.82	2.43	1.20	2.23	4.45	10.27
1937/8	8.27	3.54	1.48	3.26	2.69	10.96
1938/9	17.25	9.46	1.76	6.03	–	17.25

The Greater German Reich 1939

By 1939, with the acquisition of the Saarland, Austria, Bohemia-Moravia and Memelland, 'Greater Germany' was achieved at relatively little cost.

"Greater Germany—the dream of our fathers and grandfathers— is finally created."
Gustav Krupp von Bohlen, 1939

In 1939 the German authorities began to use the term *Grossdeutschland*, Greater Germany, when they described Hitler's Reich. The Pan-German movement in both Germany and Austria had campaigned for 50 years for the union of the German peoples of central Europe. Hitler made the creation of a single German state one of his chief ambitions in *Mein Kampf*. After the incorporation of Austria and the Sudeten areas in 1938 the only remaining German population of any size outside the Reich lay in those areas of Prussian territory which had been granted to the new Polish state after 1919. When shortly after the Munich conference the German Foreign Ministry began exploratory talks with Warsaw on the future of the city of Danzig and the Polish corridor, Hitler was taking the first step toward incorporating these areas as well into Greater Germany.

The new German state, economically revived and militarily powerful, now exerted considerable international influence. In eastern and central Europe the revival of German economic

I/The Greater German Reich, 1939
Military districts

··········	defensive line
VI	district
	army group command
	army command

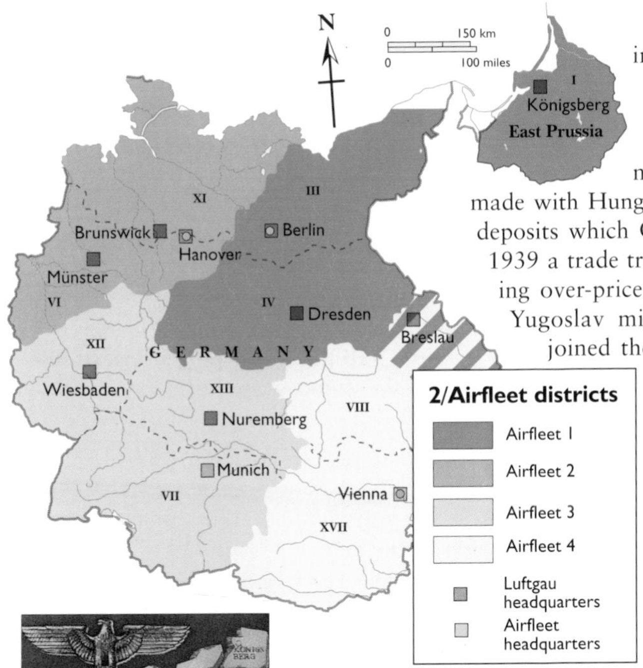
N
0 150 km
0 100 miles
I
Königsberg
East Prussia
XI
III
Brunswick
Hanover
Berlin
Münster
VI
IV
Dresden
Breslau
XII
G E R M A N Y
Wiesbaden
XIII
Nuremberg
VIII
Munich
VII
Vienna
XVII
2/Airfleet districts
Airfleet 1
Airfleet 2
Airfleet 3
Airfleet 4
Luftgau headquarters
Airfleet headquarters

interests produced a rapid withdrawal of British and French capital from a region in whose future stability they now had little confidence. In its place came German economic interests. Trade agreements were made with Hungary, which possessed large oil and bauxite deposits which Germany needed for rearmament. In early 1939 a trade treaty was signed with Yugoslavia, exchanging over-priced German armaments and machinery for Yugoslav minerals. In February 1939 Romania too joined the German economic bloc, promising supplies of oil and grain against German deliveries of arms and German investment and technical know-how in the Romanian mining industry. Slovakia, which had become a virtual satellite of the Reich in March 1939, found its economic development linked closely with German interests. The new trading bloc, centred on Berlin was christened the *Grossraumwirtschaft*, the Greater Economic Area. The region was regarded by German leaders as an economic backyard in which Germany would enjoy special privileges and access to scarce resources which she could not easily obtain from overseas.

BERLIN
13·MÄRZ 1938
EIN VOLK EIN REICH
EIN FÜHRER

A famous picture postcard of the late 1930s declares 'one people, one state, one leader'.

The areas absorbed into the Greater German Reich lost their distinct identity entirely. For Austria the loss of sovereignty brought some advantages in rising employment and production, but its status as the Ostmark in a large Germany diminished the influence and prestige of its elite. Opponents of the Nazis before the *Anschluss* were sent to the concentration camp set up in 1939 at Mauthausen. Anti-semitic legislation was applied at once, and thousands of Viennese Jews were dispossessed within months of the German takeover. The onset of Nazi repression and racism made a mockery of the plebiscite organized in Austria and Germany on 10 April 1938, when 99.07% of voters endorsed the union of the two states.

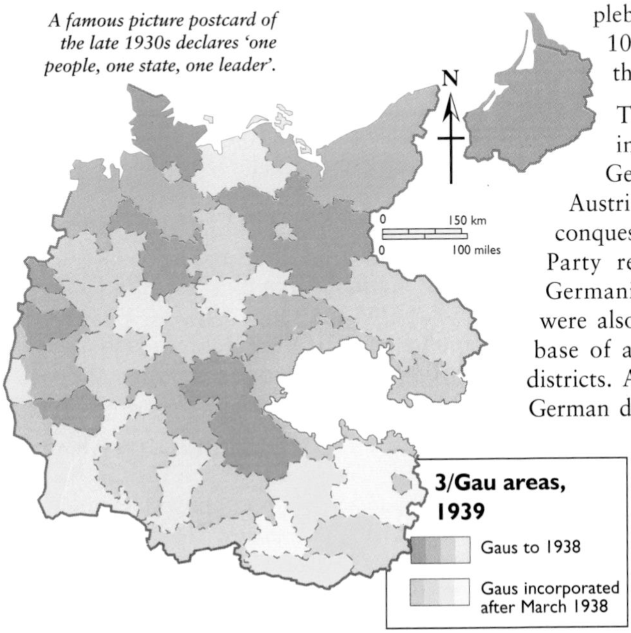
N
0 150 km
0 100 miles
3/Gau areas, 1939
Gaus to 1938
Gaus incorporated after March 1938

The whole structure of Party organization was imposed on the new regions of Greater Germany. The Sudetenland became a Gau; Austria was divided into seven Gaus. Later with the conquest of Poland the Wartheland became another Party region in which a ruthless programme of Germanization was imposed. The new populations were also liable for conscription. Austria became the base of a new air defence zone and two new army districts. Austrian conscripts were divided up between German divisions, with the provision that they should not constitute more than 25% of any one unit (Austria lost 230,000 soldiers killed during the war). The extensive Germanization of the Austrian armed forces and of Austrian public affairs created resentment among a population many of whom had longed to belong to Greater Germany before 1938.

IV: Expansion and War 1939–1945

Between 1939 and 1945 Hitler's Germany fought the largest and most destructive war in history.

"Peoples, like individuals, sometimes need regenerating by a little blood-letting"
Adolf Hitler,
August 1942

War was a defining feature of the Third Reich. Hitler always expected to fight a war of some kind and actively sought war as a solution to Germany's position in Europe. The conflict he got in 1939 turned into a long war of attrition against a combination of enemies that no-one could have anticipated in 1939. It meant that half the life of the Third Reich was spent at war. Its brief history was distorted and shaped by the military conflict, which brought about its ultimate collapse.

Hitler wanted to fight wars he could win. Against Poland in 1939 there was no question of German defeat. What Hitler had not wanted was a simultaneous conflict with the great powers. When Britain and France finally declared war on 3 September, two days after the invasion of Poland, the whole character of the war altered. The German service chiefs were much less confident of success against the western states. Hitler wanted to attack France immediately, but was dissuaded on account of the poor state of German preparation and the autumn weather, which militated against extensive air and armour operations. Of the various battle plans drawn up only that of von Manstein, a modified version of the Schlieffen Plan of 1914, appealed to Hitler. It involved a two-pronged attack, one through the Low Countries and one through the dense woods and high passes of the Ardennes forests to drive a wedge through the western line. It was a risky strategy, for if it went wrong German forces had scant reserves.

In the event, a poorly-coordinated Allied response and high operational skills on the German side brought a startling German victory in just six weeks of fighting. The victory was not quite complete, for British forces retreated back to the British mainland, and the British Government refused Hitler's unattractive offer of a German peace settlement in July. But on Continental Europe Germany was now master. Defeat of France brought a revolution in the balance-of-power. It also brought Hitler to the pinnacle of his domestic popularity.

The German domination of Europe was so unexpected that Hitler remained undecided for some months about how to exploit the changing circumstances. In July he ordered an operation against Britain in the autumn, but he was already thinking about a possible blow against the USSR with whom he had made an opportunist alliance in August 1939. As the campaign against Britain became more unlikely, with the failure to win air superiority in the Battle of Britain, so the idea of war in the east grew in Hitler's imagination. In November Germany, Italy and Japan signed a Tripartite Pact dividing the world into new spheres of influence. The USSR was invited to participate,but set demands too high for Hitler to accommodate. A few weeks afterwards Hitler signed the directive for Operation Barbarossa.

There were many motives at work in Hitler's decision. He had always viewed the east as the area for German empire. The resources he captured there could be used to defeat Britain; Soviet defeat would pre-empt any attempt by British statesmen to revive the encirclement of Germany. The east also represented the twin enemies of Germandom: a large Jewish presence and a Marxist state. Geopolitics and ideology jostled together in an uneasy embrace. In the spring of 1941 German forces were built up for a summer

MERCEDES-BENZ
FLUGMOTOREN

An advertisement for Daimler-Benz aero-engines in the magazine Luftwelt *(Air World) in 1940. At the beginning of the war Germany enjoyed a lead in most areas of aviation technology, but confused planning and the air force obsession with expensive and time-consuming refinements to existing technology undermined the technical lead.*

The German army parades through Warsaw at the end of the Polish campaign in September 1939. The battle lasted only three weeks, though the western Allies expected Poland to hold out for six months. German casualties were light. Victory in Poland was followed by the murder of much of the Polish élite, and the expulsion and transfer of millions of Polish and Jewish inhabitants.

campaign which was designed to bring Germany to the point she had reached after the Treaty of Brest-Litovsk in 1918.

Extraordinary progress was made in the first weeks of the attack, which was launched on the night of 21/22 June 1941. The date was later than Hitler wanted. Forces had been dispatched to stabilize the Balkans in the spring of 1941 following an anti-German coup in Belgrade and Italy's near failure in Greece. But there were also technical issues of preparation and logistics which made an earlier start difficult. It is now widely accepted that the sheer geographical scale of the campaign made a German victory in 1941 unlikely. By the time the autumn rains came the decisive blow had yet to be struck. At Moscow in December 1941 German forces faced their first serious reverse.

The ideological character of the war in the east produced a conflict of peculiar savagery. Well before the fighting began Hitler and Himmler indicated that German forces should ignore the usual distinction between civilian and soldier. Jews were a particular target, but Jews who were also communists were to be given no quarter. In the wake of the advancing armies came four so-called *Einsatzgruppen*, operated by the SS as hit squads against any alleged political or racial enemies of the regime. Unknowable thousands of civilians were slaughtered in the first months. Brutality spilled over into the regular army, whose view of the enemy was shaped by racist propaganda. The fanatical resistance of many Soviet soldiers, and their use of close-quarter fighting, brought high losses on the German side for the first time, and provoked brutal treatment of prisoners.

Hitler hoped to complete the defeat of the Soviet Union the following year. War with Britain intensified during 1941 in North Africa, where Field Marshal Rommel was sent to help Italian forces, and in the air war, as British bombers began the early stages of a four-year bomb campaign. Hitler now saw the resources of the Soviet south — iron, steel, oil — as the means finally to dispose of the British threat. Tension with the USA also increased during 1941 as a stream of resources crossed the Atlantic from America, and the US Navy became more embroiled in the Atlantic war against the German submarine. When Japan attacked America on 7 December 1941 Hitler and Foreign Minister Ribbentrop responded almost immediately with a declaration of war on a state they already counted as being in the enemy camp.

The declaration of war on America created an altogether different conflict. America provided economic assistance on a large scale to both Britain and the Soviet Union under the scheme of Lend-Lease. American bombers joined the RAF in what became in January 1943 the Combined Bomber Offensive. American naval power was mobilized to blunt Germany's ally

Hitler visits Paris on 28 June 1940 following the defeat of France. He arrived at five o'clock in the morning, toured most of the sights, ending with the Sacré Coeur in Montmartre. By nine o'clock he had finished. 'It was the dream of my life to be permitted to see Paris', he told the Party architect, Albert Speer.

A poster recruiting for the German submarine service, c. 1942. Submarine attack was the main activity of the German Navy during the war, and it succeeded in reducing Allied tonnage substantially in 1941 and 1942. Submariners were often absent for up to eighteen months and returned weather-beaten and bearded. Casualties were very high. Some 70% of all submariners were killed.

Japan, and to help Britain to eliminate the German threat in North Africa. German intelligence sources greatly under-estimated the speed with which America could arm, and the willingness of Americans to fight in Europe. The growing diversion of resources to combat the Anglo-American powers during 1942 made it more difficult to deliver decisive blows on the eastern front.

During 1942 German forces achieved their fullest territorial extent, reaching the Caucasus Mountains in Russia, and the Egyptian frontier in North Africa. Almost simultaneous defeats were inflicted in November, at El Alamein in North Africa and in Russia on the approaches to Stalingrad. For the next few months German forces retreated hastily in both theatres. The final attempt to reverse the tide of war came in July 1943 when Hitler launched Operation Citadel against the Kursk salient on the eastern front. The planned battle of annihilation ended with German collapse. Soviet forces had greatly improved in both technology and tactics. German forces were stretched by their vast commitments, and the stream of weapons needed to sustain modern war was never sufficient. Improvements in Allied air forces caused the large-scale diversion of manpower, guns and electro-technical equipment to the air war. In 1944 the Luftwaffe was defeated over German soil by the long-range Allied fighters.

Germany still possessed formidable power none the less. The eastern front began to solidify as German forces drew closer to the Reich. What finally unhinged the German war effort was the opening of another front in France in June 1944. After careful planning and a successful deception Allied forces captured occupied Normandy between 6 June and late July 1944. Large German forces in the West were fed piecemeal into the battle and Allied air superiority made the movement of supplies hazardous. When American armies broke out of Normandy in late July there followed a rout. The entire German army in the west was defeated and the front did not settle until Allied forces reached the German western frontier in September. There was no longer any doubt about the outcome, though Hitler clung onto the hope that new wonder weapons, such as the V2 rocket or the Me262 jet-fighter, would save Germany at the last moment. The final defeat of Germany was bitterly contested, but the relentless bombing, the collapse of organized economic life and the vast and well-equipped armies of the three main Allies could not be held at bay by sheer willpower. German forces surrendered on 7 May 1945.

Admiral Karl Dönitz in April 1945. The Commander-in-Chief of the Navy was a strong supporter of the Führer, but he was a surprise choice as Hitler's successor as head of state in the last week of the war. His belief that submarines held the key to defeating the western powers proved unfounded. Despite defeat in the Battle of the Atlantic he retained Hitler's favour.

The Onset of War 1939

In September 1939 Hitler found himself fighting the wrong war, not a war against Poland alone, but against Britain and France as well.

"I am risking this war. I have to choose between victory and annihilation."
Adolf Hitler,
November, 1939

When German troops invaded Poland on 1 September, 1939, Hitler expected a local conflict with Poland, lasting a matter of weeks. He got instead a major European war with Britain and France that lasted almost six years.

At the start of 1939 Hitler had no plans for war even against Poland. Since the Munich crisis, diplomatic pressure had been put on Poland to consider the return of the Prussian city of Danzig to the Reich, and to discuss possible readjustments to the status of the 'Polish Corridor' which separated East Prussia from the rest of Germany. In March 1939 the Polish foreign minister, Josef Beck, gave a firm refusal to all German requests. Stung by Polish intransigence, on 3 April Hitler ordered the armed forces to prepare for war against Poland. On 28 April the Polish-German Non-Aggression pact was renounced by Germany, and across the summer months German forces prepared 'Case White', the planned annihilation of Polish resistance.

Hitler intended to isolate the Poles diplomatically, and despite guarantees from Britain and France of Poland's sovereignty, he remained convinced to the end that the war could be localized. In August he sent his foreign minister, Von Ribbentrop, to Moscow to negotiate with Stalin in order to complete

I/Campaign in Poland
1–28 September 1939

German attacks

Soviet attacks

Soviet line by 20 Sept

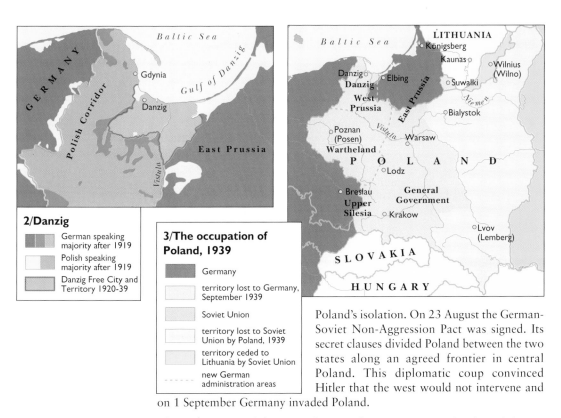

2/Danzig

- German speaking majority after 1919
- Polish speaking majority after 1919
- Danzig Free City and Territory 1920-39

3/The occupation of Poland, 1939

- Germany
- territory lost to Germany, September 1939
- Soviet Union
- territory lost to Soviet Union by Poland, 1939
- territory ceded to Lithuania by Soviet Union
- - - - new German administration areas

Poland's isolation. On 23 August the German-Soviet Non-Aggression Pact was signed. Its secret clauses divided Poland between the two states along an agreed frontier in central Poland. This diplomatic coup convinced Hitler that the west would not intervene and on 1 September Germany invaded Poland.

The German armed forces made meticulous preparations for the Polish war. They committed 52 divisions (against Poland's 30), organized into five armies surrounding Poland on three sides. Tank and motorized divisions spearheaded the attack, supported by 1,500 aircraft (against 400 Polish). Polish forces planned to fight a holding action before falling back on the defence of Warsaw. When the campaign opened German forces moved with great speed and power, quickly penetrating the defensive screen and encircling Polish troops. On 17 September Soviet forces began to cross the frontier in the east against only light resistance. Caught between the two great powers, Polish fighting power evaporated. Warsaw surrendered on 27 September. The following day all Polish resistance ceased.

The campaign cost German forces 8,082 killed and 27,278 wounded, with the loss of 285 aircraft. The whole of western Poland came under German control. On 28 September Soviet and German representatives met to draw up a demarcation line which gave Warsaw to the Germans and the Baltic states as a sphere of interest to the USSR. Almost at once the German authorities began to break Poland up. Silesia and the Corridor became parts of the Reich, and a central Polish area called the General Government was placed under a Nazi administrator, Hans Frank. Thousands of Polish intellectuals were rounded up and murdered. Polish peasants were moved from their villages in parts of western Poland and replaced by German settlers.

During the summer of 1939 German forces and arms were secretly smuggled into the Nazi-dominated city of Danzig. When war broke out on 1 September Polish policemen and frontier guards were swiftly overcome and the city was reunited with Germany.

Hitler had been right to calculate that Britain and France would give Poland little help, but he was wrong about localizing the conflict. On 3 September, two days after the invasion, Britain and France both declared war. Apart from isolated raids by scouting parties and aircraft little happened. After the defeat of Poland Hitler wanted to wage a winter campaign in the west. Bad weather prevented it, and both sides sat through a winter of 'phoney war'.

The Conquest of Europe

In the spring of 1940 Hitler began a series of campaigns which achieved German domination of Continental Europe by April 1941.

"Wherein lies the secret of this victory? Indeed, in the enormous dynamism of the Führer"
Eduard Wagner, Quartermaster–General, 1940

In the spring of 1940 Hitler was finally able to launch his attack in the West. But before the main campaign fears that Britain might launch a pre-emptive invasion of Scandinavia led Hitler to attack Denmark and Norway on 9 April. Denmark was quickly pacified, but Norwegian resistance, backed by British military aid, was kept up until early June. Both states were occupied and brought under German political control.

While the battle continued in Norway the German armed forces began the assault on France on 10 May, moving quickly into the Netherlands and Belgium in the north and the Ardennes forest further south. Using aircraft and mechanized army units as a mailed fist, German armies avoided the main French defences of the Maginot Line and drove deep holes in the enemy frontline holding northern France. The bulk of German infantry then moved forward

Conquest of Europe

- Germany 1940
- Axis 1940
- Allies 1940
- Soviet control 1940
- neutral
- German
- Allies
- Allied retreat
- Italian
- Soviet

Map labels: Narvik, Namsos, Trondheim, Andalsnes, *May 1940*, Bergen, Stavangar, Helsinki, Tallinn, Leningrad, ESTONIA, Riga, LATVIA, Moscow, Stockholm, Oslo, *Baltic Sea*, LITHUANIA, Kaunas, Smolensk, Minsk, Königsberg, Danzig, Copenhagen, DENMARK, *9 April 1940*, Hamburg, Berlin, POLAND, Warsaw, *River Vistula*, *September 1939 occupied by USSR*, Kharkov, Kiev, *annexed to Hungary 1940*, *River Elbe*, *North Sea*, Amsterdam, Rotterdam, UNITED KINGDOM, *River Rhine*, GERMANY, Prague, Vienna, Budapest, *1940*, Odess, BESSARABIA, IRELAND, London, Brussels, Cologne, Frankfurt, Munich, HUNGARY, *6 April 1940*, ROMANIA, *British withdrawal May–June 1940*, Dunkirk, Sedan, *10 May 1940*, Berchtesgaden, Belgrade, Bucharest, Paris, Châlons-sur-Marne, Bern, SWITZ., Zürich, Geneva, *River Po*, Trieste, Split, YUGOSLAVIA, BULGARIA, Sofia, *6 April 1941*, *River Seine*, FRANCE, *River Loire*, *River Rhone*, Dubrovnik, ITALY, ALBANIA, *ATLANTIC OCEAN*, Bordeaux, Marseille, CORSICA, Rome, Naples, GREECE, Athens, Corinth, *River Ebro*, SARDINIA, Barcelona, *Mediterranean*, SICILY, MALTA, SPAIN, Madrid, *♦ under Axis control after the fall of France (Vichy administration)*, Tunis, Algiers, Constantine, *Afrika Korps 14 Feb. 1941*, Misurata, Benghazi, Buerar, TUNISIA, Mareth, Tripoli, Syrte, Gibraltar, Oran, Oujda, Tangier, Fez, ALGERIA, Marble Arch, El Agheila, Casablanca, MOROCCO, Safi

Operational strengths

Army size		Tanks		Aircraft		German divisions for	
Germany	**141 divisions**	**Germany**	**2,445**	**Germany**	**4,020**	Norway/Denmark	9
Britain	10 divisions	Britain	310	Britain	1,306	Yugoslavia/Greece	24
France	104 divisions	France	3,063	France	1,368	North Africa	2
Belgium	22 divisions	Belgium	10	Belgium	250		
Netherlands	8 divisions	Netherlands	1	Netherlands	175		
Allied Total	144 divisions	Allied Total	3,384	Allied Total	3,099		

I/Operation 'Sea Lion'

I	Assembly areas
→	planned German deployments
▬	German operational objectives
▒	British minefield

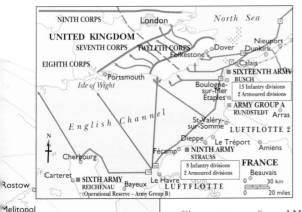

rapidly to exploit the damage done by aircraft and mobile forces. The British expeditionary force was cut off and retreated from Dunkirk back to England by early June. As the French front crumbled Marshal Pétain sought an armistice on 21 June. At Hitler's insistence it was signed in the railway car at Compiègne where Germany had sued for peace in November 1918.

The campaign lasted a mere six weeks and cost only 27,000 lives in contrast to the millions killed in the Great War. Hitler was uncertain how to proceed once the historic enemy was defeated. He made a half-hearted offer of peace to Britain on German terms in a Reichstag speech on 19 July but got no response. But even before the 'peace offer' he had ordered preparations for the invasion of Britain. The invasion, codenamed Operation Sea Lion, was planned for the autumn as long as British airpower could be neutralized. The plan was to place two armies of 100,000 men in southern England and drive on London. The failure to secure air superiority over southern England during the Battle of Britain, and problems mobilizing sufficient shipping space, forced postponement. Hitler had lost interest in invasion by August. He began thinking about a possible war against Stalin's Russia.

Circumstances forced Hitler to turn his attention to southern Europe as well. His ally Mussolini used German victories as an opportunity to extend Italian power. In October 1940 Italy invaded Greece, and Italian forces fought to capture Egypt from the British. Severe setbacks in both enterprises invited German intervention. While Greek forces threatened to defeat the Italian invasion an anti-German coup was staged in Yugoslavia. On 6 April Germany bombed Belgrade and in a matter of days had conquered the country, German forces then moved to Greece where a combined Greek-British force was defeated by late April. In North Africa two divisions were sent under Field Marshal Rommel to form the *Afrika Korps* which drove British forces back to the Egyptian frontier.

German soldiers close in on the defeated British army around Dunkirk. Delays in the attack allowed 370,000 British and French troops to reach England.

Operation Barbarossa

In June 1941 Hitler launched his greatest military gamble, the invasion of the Soviet Union. German forces proved unstoppable until they reached Moscow in December.

"The Führer estimates that the operation will take four months. I reckon on fewer. Bolshevism will collapse like a pack of cards."
Joseph Goebbels,
June 1941

On 22 June 1941 German forces, together with the allied armies of Hungary, Romania and Finland, threw three million men, 3,000 tanks and 2,700 aircraft across a thousand mile long frontier with the Soviet Union. Geographically and numerically it was the largest operation ever launched.

The decision to attack the USSR was finally made in December 1940 when 'Case Barbarossa' was laid down in Führer Directive number 21. But Hitler had seriously explored the idea since the summer. The alleged weakness of the Soviet force after the Stalinist purges presented a tempting opportunity; while Stalin's own ambitions in eastern Europe looked increasingly threatening. Above all conquest in the east promised a vast area for exploitation, the 'living space' that Hitler had talked about in *Mein Kampf*. The fact that the space was currently occupied by the ideological and racial enemies of the new Reich gave the operation the air of a crusade.

Every effort was made to conceal the preparations. The campaign of deception was designed to show that Britain was the intended target of the 1941 operation. But by the late spring there was abundant evidence of major German troop movements in the east. Stalin remained impervious to all suggestions that a German attack was imminent. Local commanders made some limited preparations, but when German forces attacked almost complete surprise was achieved. Although Soviet forces had vastly more tanks and aircraft and approximately the same number of divisions, the poor quality of much of the equipment, coupled with the disorganization of the Soviet response, created ideal opportunities for the experienced German armies to exploit.

The German allied forces attacked on three broad fronts, North, Centre and South. The first aimed towards Leningrad, which was encircled by July 1941, and subjected to a 900 day siege. The central front drove past and encircled whole Russian army groups before arriving fifty miles from Moscow in September. The southern flank drove towards the key economic targets in the industrialized southern Ukraine and by October much of the area was in German hands. Over two million Soviet prisoners were taken in three months. In October Hitler announced that victory had been achieved but his claim proved premature. Bad weather, the long and difficult supply lines, the heavy losses of men and equipment, all held up the German effort to kill off Soviet military power. In December General Zhukov organized the defence of Moscow and then a powerful counter-offensive which drove the German front back in appalling winter conditions. Both sides sat back to wait for late spring weather when fighting could begin again. Hitler remained confident that the Soviet Union could be finished off in a brief summer campaign before turning back to deal with Britain, and with a new enemy, the United States, on which Hitler had declared war on 11 December.

German tanks roll across the Soviet steppe in the summer of 1941. The great bulk of Hitler's army still relied on rail, horse and foot.

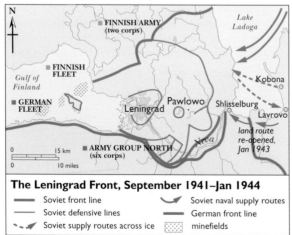

The Leningrad Front, September 1941–Jan 1944

— Soviet front line
— Soviet defensive lines
-•-•- Soviet supply routes across ice
⌄ Soviet naval supply routes
— German front line
▦ minefields

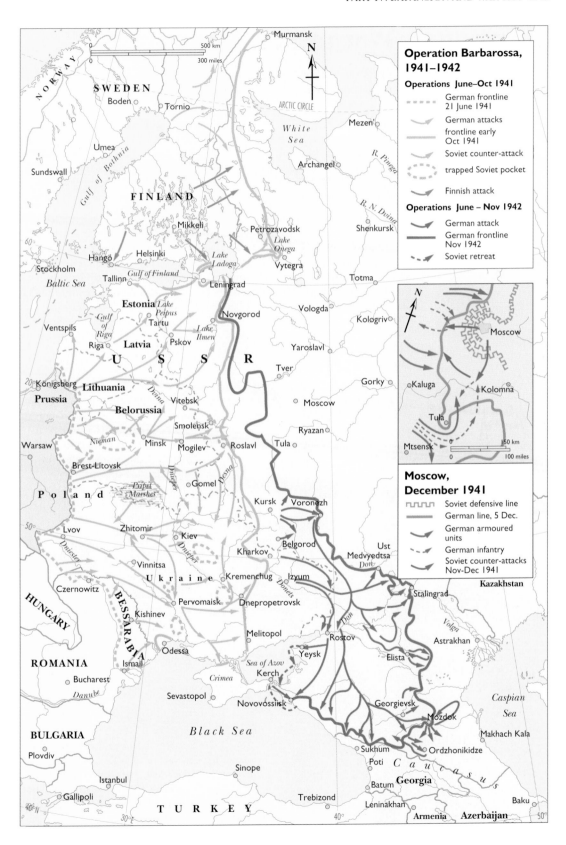

Operation Barbarossa, 1941–1942

Operations June–Oct 1941

German frontline 21 June 1941

German attacks

frontline early Oct 1941

Soviet counter-attack

trapped Soviet pocket

Finnish attack

Operations June – Nov 1942

German attack

German frontline Nov 1942

Soviet retreat

Moscow, December 1941

Soviet defensive line

German line, 5 Dec.

German armoured units

German infantry

Soviet counter-attacks Nov-Dec 1941

Germany's War at Sea

As in the First World War, German leaders gambled on knocking Britain out of the conflict by a submarine blockade.

"The sea war is the U-boat war. All has to be subordinated to this main goal."
Admiral Karl Dönitz, February, 1943

The navy was the weakest of Germany's armed services when war broke out in 1939. Against the 22 battleships and 83 cruisers of the French and British navies, Germany had only three small 'pocket' battleships and eight cruisers. The battleship *Bismarck* was launched in 1940 but was sunk on its maiden voyage in May 1941. Her sister ship *Tirpitz* was bottled up in Norwegian waters where British bombers sank her in November 1944.

Early in the war the German Navy under Admiral Erich Raeder recognized that the submarine offered the only effective German action at sea. In 1939 there were only 57 U-boats available, and not all these were suitable for the Atlantic. Under Admiral Karl Dönitz the submarine arm expanded steadily and soon took a heavy toll of Allied shipping. Dönitz introduced a new tactic to undersea warfare. The submarines, or 'wolf packs' hunted at night linked by radio. In 1940 over 1,000 Allied ships were sunk, one quarter of British tonnage. In 1941 1,299 ships were sunk and British exports fell to almost one third of the pre-war total. When the United States entered the war late in 1941 the wolf packs picked off American merchantmen still sailing singly and unescorted. In the first four months of 1942 2.6 million tons of Allied shipping was sunk.

During 1942 the submarine came closer than at any other time in either world war to undermining fatally the British war effort. U-boats operated in the Atlantic Gap, outside air cover, supplied by special vessels known as 'milch cows' which carried additional torpedoes and food. The naval B-Dienst broke British codes and directed submarines to oncoming convoys. Early in 1942 convoy losses reached exceptional levels.

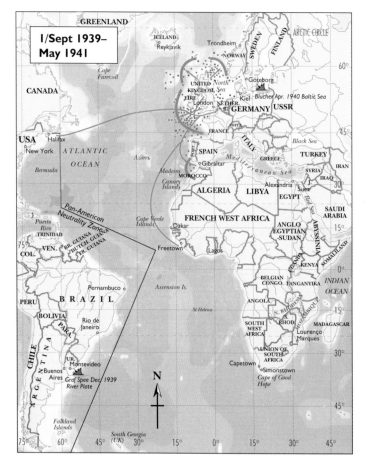

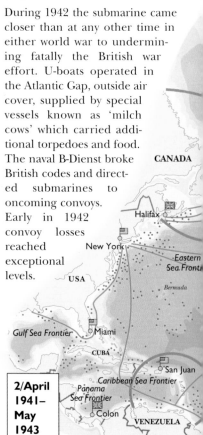

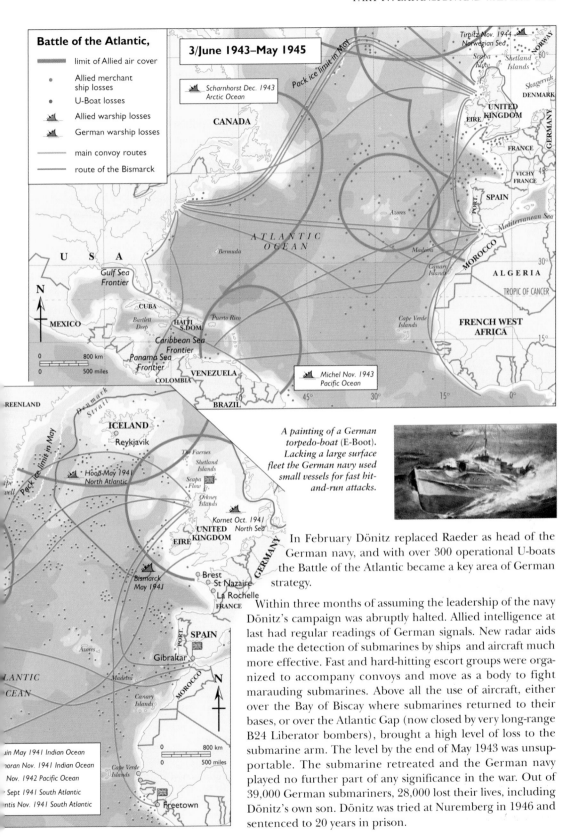

Battle of the Atlantic,

	limit of Allied air cover
•	Allied merchant ship losses
•	U-Boat losses
	Allied warship losses
	German warship losses
	main convoy routes
	route of the Bismarck

3/June 1943–May 1945

Tirpitz Nov. 1944
Norwegian Sea

Scharnhorst Dec. 1943
Arctic Ocean

Pack ice limit in May

Scapa Flow
Shetland Islands

NORWAY

DENMARK

Skagerrak

CANADA

UNITED KINGDOM

EIRE

FRANCE

VICHY FRANCE

SPAIN

PORT.

Azores

ATLANTIC OCEAN

Bermuda

Madeira

MOROCCO

Canary Islands

ALGERIA

TROPIC OF CANCER

U S A

Gulf Sea Frontier

N

MEXICO

CUBA

Bartlett Deep

HAITI
S.DOM.

Puerto Rico

Caribbean Sea Frontier

Panama Sea Frontier

COLOMBIA

VENEZUELA

Cape Verde Islands

FRENCH WEST AFRICA

Michel Nov. 1943
Pacific Ocean

0   800 km
0   500 miles

BRAZIL

REENLAND

Pack ice limit in May

Denmark Strait

ICELAND
Reykjavik

The Faeroes

Shetland Islands

Hood May 1941
North Atlantic

Scapa Flow

Orkney Islands

Kornet Oct. 1941
North Sea

pe
well

UNITED KINGDOM

EIRE

GERMANY

Bismarck
May 1941

Brest
St Nazaire
La Rochelle

FRANCE

LANTIC
CEAN

Azores

Madeira

PORT.

SPAIN

Gibraltar

MOROCCO

Canary Islands

N

0   800 km
0   500 miles

Cape Verde Islands

Freetown

uin May 1941 Indian Ocean

horan Nov. 1941 Indian Ocean

Nov. 1942 Pacific Ocean

Sept 1941 South Atlantic

ntis Nov. 1941 South Atlantic

A painting of a German torpedo-boat (E-Boot). Lacking a large surface fleet the German navy used small vessels for fast hit-and-run attacks.

In February Dönitz replaced Raeder as head of the German navy, and with over 300 operational U-boats the Battle of the Atlantic became a key area of German strategy.

Within three months of assuming the leadership of the navy Dönitz's campaign was abruptly halted. Allied intelligence at last had regular readings of German signals. New radar aids made the detection of submarines by ships and aircraft much more effective. Fast and hard-hitting escort groups were organized to accompany convoys and move as a body to fight marauding submarines. Above all the use of aircraft, either over the Bay of Biscay where submarines returned to their bases, or over the Atlantic Gap (now closed by very long-range B24 Liberator bombers), brought a high level of loss to the submarine arm. The level by the end of May 1943 was unsupportable. The submarine retreated and the German navy played no further part of any significance in the war. Out of 39,000 German submariners, 28,000 lost their lives, including Dönitz's own son. Dönitz was tried at Nuremberg in 1946 and sentenced to 20 years in prison.

The Turning of the Tide, 1942–1944

From the autumn of 1942 German forces faced their first real defeats of the war. During 1943 they were forced to retreat on all fronts, though Germany still possessed formidable military strength.

"I have terrible troubles. I see the situation in the east every day, and it is terrible…But I have always been worried about the west."

Adolf Hitler, December, 1943

During 1942 the main weight of German military effort was concentrated on the Eastern Front where Hitler insisted on an attack on the southern axis against the economic targets, particularly the oil of the Caucasus. In June 1942 German forces launched 'Operation Blue', clearing the Crimea and the remaining areas of the southern Ukraine, before dividing for attacks against Stalingrad and against the Caucasus oilfields around Maikop, Grosny and Baku. At first Axis forces made remarkable gains. By September the Red Army had been driven back into Stalingrad, which was expected to fall in a matter of days.

This was the limit of the German advance. Fierce Soviet resistance in Stalingrad, and the extended supply lines, slowed the progress of General Paulus's 6th Army to a crawl. Overstretched forces in the Caucasus were halted and then slowly driven back by Soviet forces whose supply position improved steadily as Soviet factories made good the losses of 1941. In November a powerful Soviet counter-offensive cut off Stalingrad and drove the German line back. On 31 January Paulus surrendered. Since November, 32 Axis divisions had been destroyed. In the summer of 1943 Hitler ordered one last attempt to break the Soviet front. 'Operation Citadel' drew almost one million men and 2,700 tanks to a narrow front around the city of Kursk where German

Below: *German motorcycle transport in the western desert. Rommel's small Afrika Korps drove the British back into Egypt until his defeat at El Alamein in November 1942.*

Left: *A picture from the magazine* Die Wehrmacht *showing German soldiers fighting heroically at Stalingrad, where the German army suffered its heaviest defeat since the war began.*

German divisions, summer 1944	
Eastern front	156
Western front	54
Italy	27
Balkans	20
Norway	9

German air forces, spring 1944	
Eastern Front	1,710
Western Front	1,410
inside Reich	1,225
Med & Balkans	505

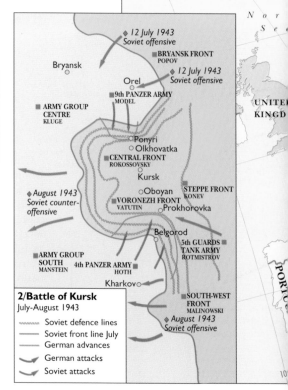

2/Battle of Kursk
July–August 1943

- 〰〰 Soviet defence lines
- —— Soviet front line July
- —— German advances
- ➤ German attacks
- ➤ Soviet attacks

military leaders planned to encircle and annihilate a Soviet force of more than one million. Careful Soviet preparations blunted the German attack between 5–13 July and there then followed a massive Soviet counter-thrust which broke the German line and in four months drove it back to Kiev. Although the German army fought a fierce defence, each successive Soviet attack forced a further withdrawal.

Meanwhile, in the Mediterranean theatre, Rommel's *Afrika Korps*, hampered by attacks on the overseas supply lines was defeated in the battle of Alamein, fought between 24 October and 4 November 1942. On 8 November American and British forces landed in Morocco and Algeria (Operation Torch) and the two allied armies bottled up Axis forces in Tunisia where they surrendered on 13 May. In July Allied forces landed in Sicily, and in September they crossed in strength into mainland Italy. Here under the effective leadership of Field Marshal Kesselring, German forces fought another bitterly contested retreat the whole length of the peninsula. By the end of 1943 German forces were fighting retreats in the east and south; with increasing partisan insurrection in the Balkans and France, the tide of German success had turned.

3/Stalingrad
Sept 1942-Jan 1943

- – – – – frontline Nov 19
- ——— frontline Nov 30
- ➤ German attacks
- ➤ Soviet attacks

◆ 19 Nov. 1942
Soviet counter-offensive
■ 8th ITALIAN ARMY
3rd ROMANIAN ARMY
4th ROMANIAN ARMY
Kalach
Stalingrad
6th ARMY PAULUS
Dec. 1942 ◆ German counter-thrust
Kotelnikivo
■ 4th PANZER ARMY

I/Turn of the tide, 1942–43

■	Greater Germany
■	occupied by Germany
■	occupied by Italy
■	axis co-belligerents
■	allied territory
□	neutral
——	borders in 1942
- - - -	Front line July 1943
✳	major action
➤	German attacks
➤	Soviet attacks
➤	British attacks
➤	US attacks

Germany's War in the Air

The great expectations of German air power were disappointed during the war. By summer 1944 the **Luftwaffe** *was no longer a significant factor in Germany's war effort.*

> *"I can only win this war if I destroy more of the enemy's cities than he destroys of ours."*
> Adolf Hitler, July 1943

During the 1930s the threat of the German air force was a central feature in the growing fear of Nazi Germany abroad. When war broke out populations in London and Paris expected an immediate and devastating attack from German bombers. None came. In the early stages of the war the air force was confined to a tactical role. In combined-arms operations in Poland, Norway and France it proved highly effective.

The first real test of the air force as an independent service came in August 1940 when it was sent against the RAF over southern England as a preparation for German invasion. The subsequent Battle of Britain, which lasted until late September, showed the limits of German air power. Although it inflicted high losses on the RAF the *Luftwaffe* lost the battle for air superiority. Between July and October 1940 it lost 1,733 aircraft and was unable to make up those numbers with new output. In early August the German fighter force had some 700 operational fighters; by 1 October there were only 275, against a British figure of 732. Frustrated at the lack of success Hitler ordered a switch in mid-September away from RAF targets to the bombing of British cities to terrorize the population into seeking peace. The so-called bombing 'Blitz' that followed killed 42,000 city-dwellers but had little effect on production and failed to dent British determination to fight on. As the air force prepared for Barbarossa the air war over Britain petered out.

1/Battle of Britain, July–October 1940

- ▣ Fighter Command Group HQ
- ▲ sector airfield
- — RAF group boundary
- ☢ high level radar station
- ☢ low level radar station
- ✈ balloon barrage
- ◗◖ observer corps centre station
- ∠ anti-aircraft artillery (with number of guns)

- ▣ Luftflotte HQ
- — Luftflotte boundary
- --- Fliegerkorps boundary
- VIII Fliegerkorps
- ⋯⋯ limit of German fighter range
- ↜ main direction of German air attacks on Adler Tag (Eagle Day), Aug 13

2/Cities heavily bombed by Luftwaffe

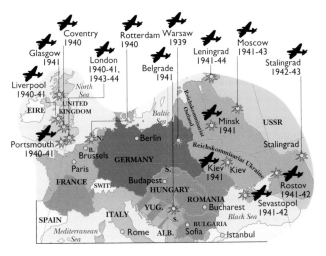

From 1941 onwards the German air force concentrated on tactical aviation. No serious attempt was made to mount a long-range bombing offensive, though a brief 'Baby Blitz' (Operation Steinbock) was launched against London in the winter of 1943–4. On the eastern front aircraft supported the army offensives and bombed objectives in front of the army advance. Even these activities had to be reduced when the bombing of Germany became a serious threat. The RAF campaign forced a major re-think of German air strategy. In 1940–41 the 'Kammhuber Line' of radar, aircraft and searchlights was set up across northern Germany. Over the next three years more and more of German air strength had to be diverted to defence. This weakened the tactical air forces which lost the initiative on the eastern and southern front and played little part in the campaign against Allied forces in Normandy. When Allied long-range fighters began to penetrate far into Germany in the spring of 1944 the air force found itself unable to perform any role effectively: The fighter force suffered losses of 25% a month and from then on the *Luftwaffe* played little part in the final defence of the Reich.

The Dornier Do 17 medium bomber (below) was one of the mainstays of the German bomber force early in the war.

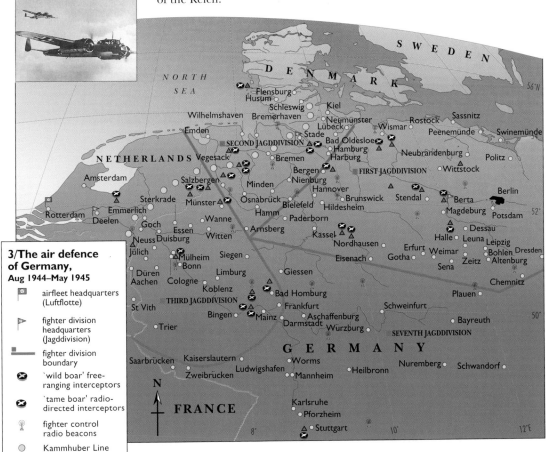

3/The air defence of Germany, Aug 1944–May 1945

- airfleet headquarters (Luftflotte)
- fighter division headquarters (Jagddivision)
- fighter division boundary
- 'wild boar' free-ranging interceptors
- 'tame boar' radio-directed interceptors
- fighter control radio beacons
- Kammhuber Line

The Defeat of the Reich, 1944–1945

Germany's fate was sealed following the western invasion of France in June 1944. Hitler would accept no surrender and died in his Berlin bunker in April 1945 as Soviet forces approached.

"If we should lose this war it will mean that we have been defeated by the Jews."
Adolf Hitler, February 1945

Throughout 1944 and early 1945 Hitler continued to argue that the war was not lost. When President Roosevelt died in April 1945 he grasped the final straw of possible Allied dissension and sent a circular to his troops to prepare for the final victory.

By 1945 Hitler lived in a world of delusion. During 1944 Germany's enemies grew in fighting skill and material power to an extent that Germany could not match. In June 1944 American, British Commonwealth and French troops invaded northern France and in three months inflicted a massive defeat on the whole German army in the west. By September they had reached the frontiers of the Reich, where they paused to regroup and re-equip for long enough for German forces to prepare a final defence of the homeland. In December Hitler gathered all the divisions and aircraft he could spare and launched a second attack through the Ardennes at a weak link in the Allied front. The so-called Battle of the Bulge was repulsed in three weeeks and laid Germany open to the final assault. In the east a series of blows from Soviet forces annihilated whole German army groups in the second half of 1944. By August the Red Army had knocked Romania and Bulgaria out of the war; by December it had reached Yugoslavia and Hungary and East Prussia.

For the final assault on Germany the western Allies had 85 divisions, 23 of

I/Liberation January 1944–March 1945

- Axis or occupied territory liberated to March 1945
- under German/Axis control, March 1945
- ✳ site of battle
- *Dec 1941* date of battle
- ↙ Allied attack

which were armoured, against a defending force of 26 divisions. In early March the German front in the west cracked. In the east the Soviet forces fielded over 14,000 tanks against 4,800 German, and put 15,500 aircraft in the air against 1,500. A series of carefully organized military punches brought Soviet armies to within 35 miles of Berlin by February. On 31 March Stalin ordered the assault on the capital. Hitler shot himself as Soviet forces approached the bunker under the chancellery on 30 April 1945. Limited resistance continued until remaining German forces surrendered on 7 May.

Right: An aerial photograph of the German port of Bremen in April 1945. The German seaboard cities were heavily hit during the war because RAF and USAAF bombers could reach them more easily amd safely than more distant overland targets. The attack on submarine production in the port cities was countered by the dispersal of submarine construction and new techniques of pre-fabrication.

2/Final Operations, April–May

- Soviet front line, 19 April
- Soviet front line, 7 May
- Soviet attacks to 7 May

- US front line, 19 April
- US front line, 7 May
- US attacks to 19 April
- US attacks to 7 May

- British front line, 19 April
- British front line, 7 May
- British attacks to 7 May

- German occupied area, May 1945
- Occupied by US, British and French forces
- Occupied by Soviet forces

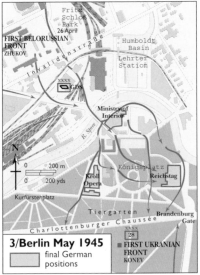

3/Berlin May 1945

- final German positions

FIRST UKRANIAN FRONT KONEV

V: The German New Order

During the war Germany became briefly an imperial state again, exploiting its conquests and pursuing a crude ethnic cleansing.

"first: conquer
second: rule
third: exploit"
Adolf Hitler,
July 1941

German rule in Europe during the Second World War reached its fullest extent late in 1942 when German forces were still pressing to seize Stalingrad on the river Volga, and an occupying force was sent into Vichy France to forestall further Allied landings following the invasion of North Africa in November. This German empire, the New Order (*Neuordnung*) as Nazi leaders called it, was an incoherent political structure held together only by German military power. Like Napoleon's Europe more than a century before its entire *raison d'être* was war. With German defeat in 1945 not a vestige of that empire survived.

The organizing principles of the New Order emerged in a haphazard way following Germany's early victories. It was formally announced on 3 October 1941 by Hitler himself when he returned to Berlin from the front to tell his people that Soviet Russia was on the point of defeat and that the work of rebuilding Europe was about to begin. The Soviet Union was not defeated in 1941, but the process of reconstructing Europe economically, politically and racially was set in motion as far as it was possible under conditions of continuous warfare.

Of all the elements of the New Order the economic was the most conceptually advanced. Even before the outbreak of war German foreign economic policy was directed to trying to create a wider economic bloc in Europe which would be dominated by German economic interests and would swing the focal point of the European economy away from Paris and London. The defeat of France and the exclusion of Britain from the Continent in 1940 cleared the way. The German Economics Minister, Walther Funk, launched in July 1940 a concerted attempt to bring the trade and financial affairs of Europe into a single system centred on Berlin, in which the Reichsmark would become the major reserve currency and continental trade and international payments would be monitored by German authorities in a German 'common market'. Such a vision was not entirely rejected by the business communities of the conquered areas, where there had been growing interest before 1939 in a more rational and regulated market structure. The reality proved very different from the vision, however. German interests were put first; countries that traded with Germany were forced to accept blocked marked accounts as payments, and in effect to extend generous credits to Germany over the war period. German businesses bought up foreign producers at cheap cost, or forced them to join European cartels dominated by German firms. Much of European heavy industry came under the direct management of the German state-owned Reichswerke, which exploited them entirely for the needs of the German war economy. The labour market in the occupied areas was also distorted by German recruitment of voluntary and forced workers, more than seven million of whom worked in Germany by 1944, and millions more in firms working for Germany outside its frontiers.

A German anti-Bolshevik poster. During the war the regime used this propaganda to win the support of anti-communists in other European states. As the war turned sour so Goebbels tried to change the image of German war-making into a crusade against Marxsim in defence of European civilization.

The political organization of Europe was more complex. German leaders did not want to rule the whole of Europe directly, since they lacked the resources and experience to do so. In the north and west of the continent the running of local affairs remained in the hands of local authorities whose work was

A Dutch anti-Nazi cartoon shows Hitler, Göring and Goebbels as the harbingers of death and destruction. Hitlerism was presented in Allied propaganda and in resistance writing as the embodiment of human evil, while the anti-Hitler cause was seen as both moral and just. The war was a struggle for belief as well as power.

monitored by German commissioners (Netherlands, Norway, Denmark) or by military governors (Belgium, France). In the east and south-east, large parts of which were destined to be the core of a German Eurasian empire. German rule was more direct. Austria, Bohemia and much of conquered Poland came directly under Berlin, though Bohemia remained technically a Protectorate. In the Balkan region the German situation was complicated by the presence of Italian occupation forces, and the survival of large-scale partisan activity, which made the Balkans an area of military action long after the formal defeat of Yugoslavia and Greece. Yugoslavia found itself in the unique position of experiencing on its territory all the different forms of German rule simultaneously. Northern Slovenia was absorbed into the Reich as an area deemed capable of Germanization. Croatia, a satellite of Germany under the Ustasa regime, was to be given independent status; Serbia was placed under German military authority in the person of General von Schröder, with separate offices for military and civilian questions, and an economic plenipotentiary who took orders direct from Göring as head of the Four Year Plan. The confused nature of authority in Serbia was compounded with the decision to appoint a native Serbian government to conduct local affairs under General Milan Nedic.

An election poster for the Nazi puppet premier of Norway, Vidkun Quisling. He founded the Nasjonal Samling in 1933. He believed that Norway's future lay with an alignment with Hitler's Germany. In February 1942 he was appointed prime minister. In 1945 he was tried for treason and shot.

als Luftnachrichtenhelferin

A wartime recruiting poster for women to join the airforce auxiliaries. Women helped to man searchlights and anti aircraft posts, and acted as aircraft observers. All three armed services used women in administrative roles. The view that women in Germany did not work for the armed services is a myth.

Affairs in other parts of the new Empire were scarcely less complicated. In the Ukraine, which some Nazi leaders, notably Alfred Rosenberg, the Party's official ideologue, wanted to turn into a new national state friendly to Germany, local Party bosses saw it as an area for crude German colonization. The Reichskommissariat Ukraine was established on 20 August 1941, and although parts of the region were still being fought over, Hitler placed it under Gauleiter Erich Koch, a civilian and an old Party fighter. He applied a harsh colonial rule, extorting economic resources, disregarding even the most basic Ukrainian interests. In practice many Ukrainians remained in junior administrative positions for want of any alternative, but they remained thoroughly alienated from the German commissariat, nullifying any hope that they might have welcomed the German New Order as a preferable alternative to Stalinism.

The third area of New Order activity was race. German conquest made it possible to export the biological politics of the Reich to the rest of Europe. This included the search for, and kidnapping of, children deemed to have the necessary 'aryan' features, to be brought up in the Reich, and the liquidation, through murder or neglect, of psychiatric patients and the mentally and physically disabled. But at its core was the opportunity to do something about the so-called 'Jewish question'. The regime's Jewish policy went through a number of stages after the outbreak of war. Hitler hoped for some form of compulsory emigration, perhaps to the French island of Madagascar, which he saw as a potential tropical ghetto where Jews would die of disease and malnutrition. While this option was not entirely closed in 1940 and 1941, the authorities began a programme of ghetto-building in Europe itself, with hundreds of thousands of Jews transported across Europe to ghettoes in the east, where Jewish councils administered them in uneasy collaboration with the German authorities. The invasion of the USSR changed the circumstances again. Millions of Jews in the areas targeted for conquest required a definite decision on the racial future of the new empire. Before the invasion Hitler ordered harsh measures against 'Jewish-Bolsheviks', and thousands of Jews were openly murdered and thrown into mass graves throughout the Baltic states, Belarus and the Ukraine.

Although the precise dates at which the key decisions in the Jewish question were taken have yet to be established, the evidence points strongly to a decision in July, as German forces won startling victories in the USSR and Hitler came to realize that victory would give him the freedom to carry out whatever policy he chose on the race question. Göring ordered Heydrich in July to work out a 'Final Solution', and Adolf Eichmann, who was in charge of the rail transport of Jews in Europe, later recalled that Himmler told him in July that 'physical extermination' was what Hitler wanted. By the autumn the murder of Jews was routine. Experiments with mobile killing vans, in which chemical agents were used to kill Jews and mental patients packed naked into the windowless trailers, convinced the authorities that a less conspicuous and more rational form of extermination should be set up, based on camps designed like factories to process human beings. Building began in 1941 and the extermination camp at Auschwitz-Birkenau was completed in October. In March 1942 the SS launched Operation Reinhard to liquidate Poland's Jews at three further camps, Treblinke, Sobibor and Belzec. Over the next three years an estimated five or six million Jews were killed, brought from all over occupied Europe, and from the satellite states, such as Croatia, which collaborated willingly on the Jewish genocide.

The savagery and ruthlessness of much German rule in Europe provoked all kinds of resistance activity, from minor acts of dissent to full-scale partisan war. But resistance was by no means universal. In all the occupied areas the German authorities found groups of collaborators, from native fascist organizations, such as the Hungarian Arrow Cross or the Belgian Rexists, to business leaders who needed full order books. For many Europeans, subject to German rule for three or four years, some degree of normality was restored to daily life after the initial crisis of occupation. Black-marketeering and corruption produced large wartime profits. Jewish shops and housing were taken over by local businessmen. German rule also gave the opportunity for other political movements to prosper with German acquiescence. In the east anti-communist Russians were organized into an 'army' under General Vlasov (1900–1946), who was captured in July 1942 and agreed to launch a Russian Liberation Movement. His army consisted largely of Russian POWs who were keen to avoid imprisonment. Vlasov hoped to restore a Russian state after a German victory, and in this disagreed with Hitler's view of the Russian future. His force gave limited assistance to the German army, and fought in the final battles around Prague in 1945.

Vlasov, like a great many collaborators, was hanged after the war. Across liberated Europe revenge was taken on all those who had profited from or collaborated with the German occupiers. Many German military and political officials were sent for trial in the occupied areas for specific crimes committed in the areas they had ruled. Fascist movements all over Europe collapsed with German defeat. The New Order was a front for German economic exploitation, political dominance, and racial engineering. It was created by violence thrived on violence and was violently destroyed.

A Waffen-SS recruiting poster from April 1944 encouraging Flemings to join the campaign against 'Jewish-Bolshevism'. During the war thousands of young European men responded to the call.

German Rule in Europe

While fighting a war on three fronts the Nazi regime began the political restructuring of Europe in German interests.

"The aim of our struggle must be to create a unified Europe."
Joseph Goebbels, 1943

German administration of the New Order created by its military successes took a variety of forms. The areas taken over before the war were integrated into the Greater German Reich. So too were areas captured after September 1939 which had been part of the pre-1918 Reich–Alsace Lorraine, Silesia and the Polish Corridor through Prussia. All of this territory was to form the core of the German empire in Europe. Beyond the frontiers of this new Germany were two distinct types of administration: areas under German control with a civilian administration and areas under the direct authority of the military occupation forces.

Military control was maintained in areas which were regarded as theatres of war in the captured areas of the USSR, in northern France and Belgium, in parts of the Balkans and, from 1943, in Italy. In other captured areas civilian administration was maintained, either by the appointment of a German ruler, as in Poland or Bohemia-Moravia, or through the establishment of a pro-German regime, as in Norway or Croatia. Nazi Gauleiters were appointed for the areas integrated with the Reich; in the other regions a Nazi commissioner represented German interests alongside the military or native

I/German administration

– – – – borders in Nov 1942

■ German civil administration

▲ German military administration

Greater Germany

occupied by Germany

occupied by Italy

Co-belligerents

Allied territory

neutral

2/Yugoslavia, Albania, Greece

— boundaries before Axis invasion

under German civil administration

under German military administration

Croatia

to Italy

to Albania

to Hungary

to Bulgaria

civilian administration. Some Nazi office-holders—Hans Frank in the Polish General Government, Erich Koch in the Ukraine, Artur Seyss-Inquart in the Netherlands—established small political empires of their own, emulating the savage state terror and racism of the Reich itself.

In some areas the political fate of conquered states was left to the period after the end of the war. In France a hybrid form of control was established following the armistice in June 1940. Northern France came under direct German military rule, with a military governor in Paris; the southern part was allowed a semi-independent existence with a government at Vichy under the veteran soldier Marshal Pétain. The Vichy regime remained responsible for organizing the daily life of the northern part, and for paying the salaries of teachers and officials, but it had no sovereignty in the area. On 11 November 1942, in reaction to Allied landings in North Africa, the rest of France was occupied by German and Italian forces. The southern area was ruled by a second military occupation authority based on Lyon, with a military commissioner, General von Neubronn, as the liaison between Vichy and the two occupation zones. The result here, as in much of occupied Europe, was an extraordinary jumble of authorities whose responsibilities were difficult to untangle. There were endless arguments across occupied Europe between military authorities, Party representatives and local politicians and officials. The only effective instrument of co-ordination was the RSHA, the only German agency to operate a single centralized apparatus covering the whole of the occupied area. The German security services enjoyed special powers which they exerted to stamp out political dissent and resistance in New Order Europe.

In all the occupied areas effective local administration depended on the recruitment of collaborating groups who either willingly co-operated from ideological conviction or self-interest or did so in order to protect local populations from the worst effects of occupation. German rule of most of Continental Europe would have been impossible without collaboration.

Polish Jews performing forced labour duty near Warsaw shortly after the German occupation. Poland was partitioned and most of its Jewish population murdered.

3/France

— boundary of France before May 1940

annexed to Greater Germany

under German military administration

closed military zone

independent French state

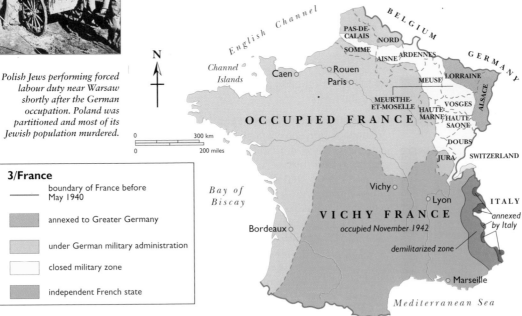

Planning the Post-War Order

With the campaign in the USSR Hitler began to plan the shape of the post-war Germanic order from the Atlantic to the Urals.

"Now we have the Russian spaces.. We'll get our hands on the finest land, and we'll guarantee for ourselves the control of the vital points. We'll know how to keep the population in order."
Adolf Hitler,
February 1942

The conquest of Europe opened up to Hitler and his Party élite the prospect of remodelling the whole European order once peace was established. The process of planning and preparing the new empire began almost as soon as the first military triumphs were achieved.

The plans for the new Europe were racial, political and economic. German planners saw Europe in terms of a strict hierarchy of races. A more privileged place was to be accorded the 'Nordic' peoples of Scandinavia and the Low Countries. The slav peoples of the east were to be treated as inferior beings, fit only to labour for the new master race (*Herrenvolk*). The Latin and Balkan peoples of southern Europe had an ambiguous place in German long-term plans, as allies but not as equals of the Nordic races.

The racial division of Europe matched closely the planned economic structure. German plans envisaged a kind of involuntary Common Market in which a single trading and currency bloc would function with the Reichsmark as the reserve currency and Vienna and Berlin as the twin financial and commerical capitals of the new system. The process of centralizing Continental finance and trade on Berlin began as early as 1940. Individual German businesses—most notably the chemical giant I.G. Farben—began to re-order European commerce and industry into cartels and trade associations centred on the Reich. The giant Reichswerke 'Hermann Göring', a state-owned holding company, took over most of the captured heavy industry and its directors planned a massive programme of industrial development stretching from central Germany to the Donetz Basin in the Ukraine in order to shift the main weight of European industry to the Eurasian heartland ruled by Germany. The whole region was to be bound together by a continent-wide network of *Autobahnen*, a wide-track railway to speed the new imperial rulers deep into Russia, and an Adolf Hitler-Canal linking the Oder, Elbe and Danube rivers.

The political settlement of the new Europe remained unresolved. Some kind of limited sovereignty was anticipated for the areas of western and northern Europe. The only region where definite political plans were drawn up was the east, but even here Hitler refused to make a final decision and there existed a number of competing conceptions of the imperial future. Hitler's own preference was for the annexation of the Baltic states to the Reich, and the establishment of a vast area of German settlement and colonization in western Russia and Ukraine, with the Ural

Hitler's new chancellery in Berlin, built by Speer in the 1930s, was to be the heart of the new continental empire, where Germany's subjects would do homage.

mountains as the putative frontier between Germandom and Slavdom. Beyond the Urals would exist a feeble agrarianized state inhabited by all the surviving Slavic elements not needed for forced labour. Alfred Rosenberg, the man Hitler appointed as Minister for the Eastern Area, had rather different ideas. He wanted to develop pro-German national states in the east and to present German invasion as liberation. The Rosenberg Plan for the east did produce civilian governments in the Baltic States and Ukraine, but both areas quickly became provinces of a Nazi empire, subject to ruthless Germanization. Liberation was promoted as a political tactic. Native interests were ignored and during 1942 the first train-loads of German colonists made their way eastward to an uncertain future.

I/Nazi Plans for Europe

- Greater Germany
- territory to be annexed to Germany
- independent Asian states under German supervision
- proposed Russian national territory
- planned allied/ satellite states
- neutral states
- ----- planned Eurasian wide-gauge rail network

Allies, Collaborators and Imitators

All over Europe German victory brought other fascist movements into positions of power, keen to imitate the German example.

"The new Norway must build on Germanic priciples, on a Norwegian and a Nordic foundation"
Vidkun Quisling, 1942

The Nazi movement in the 1930s was part of a broader European political shift to the extreme right. During the war Germany was able to exploit this development by supporting puppet fascist governments in occupied areas, (such as Norway) or winning the active collaboration of fascist or quasi-fascist regimes. The powerful anti-Marxism that fuelled the rise of the extreme right also produced thousands of volunteers from other European states to fight against the Soviet enemy.

Nazi Germany's chief ally was Mussolini's Fascist Italy. In May 1939 the two regimes entered into the so-called Pact of Steel, a military alliance from which Mussolini extricated himself with some difficulty in September 1939. In June 1940 he joined in the war on Germany's side, and in the Balkans and North Africa he did get German military co-operation. Although Hitler kept the *Barbarossa* campaign secret from his fellow dictator, Mussolini sent forces to help in the east in 1941. So too did Franco. In October 1941 a division of volunteers from the Spanish fascist Falange party went to Russia. It was withdrawn in 1943 but a volunteer Spanish Legion stayed on, and was among the final forces defending Berlin in April 1945. Other volunteers came from all over Europe, many of them attracted into the elite SS which recruited hon-

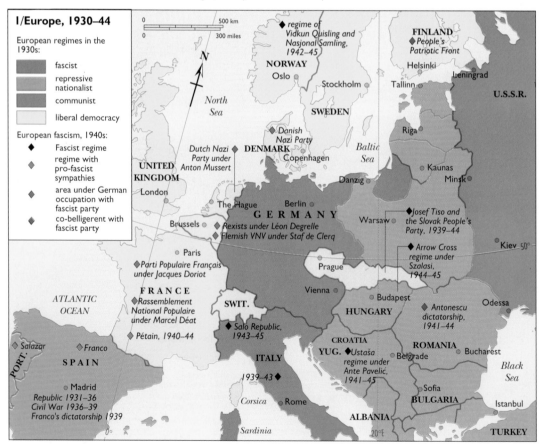

I/Europe, 1930–44

European regimes in the 1930s:

- fascist
- repressive nationalist
- communist
- liberal democracy

European fascism, 1940s:

- ◆ Fascist regime
- ◆ regime with pro-fascist sympathies
- ◆ area under German occupation with fascist party
- ◆ co-belligerent with fascist party

orary 'aryans' for combat in the armed SS divisions (Waffen-SS) formed in late 1939. More than 200,000 volunteers joined during the war, including a small British Free Corps.

Under German pressure much of Europe had fascist or pro-fascist regimes during the war. In Slovakia the clerical-fascist Slovak People's Party under Josef Tiso was installed in power in 1939 and was entirely subordinate to Berlin. Some 50,000 Slovaks served on the eastern front, but increasingly Slovak nationalists became disillusioned with Nazism and an anti-German rising was staged in August 1944 which was bloodily suppressed. Other allies in Hungary, Romania and Bulgaria, all of which sent forces to fight in the east, became lukewarm about the German cause as the war went on. Romania changed sides abruptly when Soviet forces reached her borders in August 1944, and Hungary under Admiral Horthy was occupied by German forces in March 1944 and a pro-German regime installed. Italy, too, dropped out of the alliance in September 1943, and became a German-occupied state. In northern Italy a new Fascist state, the so-called Salo Republic, was set up, and Fascist forces continued to fight on Germany's side up to the end of the war.

Above: One of many posters calling on non-Germans to volunteer for Germany's fight against Communism. This one was targeted at the Belgian Walloon community.

Right: General Francisco Franco at the end of the Spanish civil war, in 1939. Franco received aid from Germany and Italy and was sympathetic to their cause during the war, though Spain remained neutral.

2/Axis volunteers

Greater Germany

occupied by Germany

occupied by Italy

Axis co-belligerents

Allied territory

neutral

volunteer unit for Eastern Front

local German volunteers for Waffen-SS

volunteer corps for Waffen-SS

total of foreign Waffen-SS recruits

0 — 500 km
0 — 300 miles

FINLAND 1,200

6,000
1,218

NORWAY 300
Oslo
Helsinki
Leningrad
Tallinn 20,000

SWEDEN
Stockholm

1,292
1,164

Riga 31,450

Cossacks c. 30,000
Caucasian 25,000
Turkestanis 1,200
Azerbaijanis 20,000

U.S.S.R.

6,000

DENMARK
Copenhagen
Baltic Sea

Reichskommissariat Ostland
Kaunas

UNITED KINGDOM c. 55,000
London 2,559

Danzig
Minsk 20,000

The Hague 875
Brussels
GERMANY Warsaw

Reichskommissariat Ukraine

Kiev 50°

23,000 Flemings
20,000 Wallons
Paris

122,000 conscripts from areas incorporated into Greater Germany

20,000
FRANCE
Prague
5,390

Vichy
LIECH. 80
SWIT. 800
Vienna
Budapest

22,125
Odessa

HUNGARY 54,000
396 inf. div.
369 inf. div.
373 inf. div.
392 inf. div.

ROMANIA
Belgrade
Bucharest

Black Sea

ATLANTIC OCEAN

Legion tricolore

ITALY
YUG.
17,538
Split

21,516
Sofia

PORT.

SPAIN

Madrid

250 inf. div.

Rome
ALBANIA
BULGARIA
Istanbul

20°E

TURKEY

North Sea

N

87

Exploitation and Plunder

German conquests in Europe brought rich economic rewards. A flow of labour, food and money fuelled Germany's own war effort.

"As for myself, I think of looting, comprehensively."
Hermann
Göring, 1942

Economics stood at the heart of the German New Order. The areas of conquered Europe were regarded as an additional resource for the German war effort, providing the manpower and resources lacking within Germany, as well as food and finance. Without these extra productive resources the German war effort would have been plagued by the same crises it faced during the First World War.

Economic exploitation varied from area to area, and it changed over time. In the areas taken over in the late 1930s industry and agriculture were closely tied to the German economy. The Four Year Plan played a major part in organizing the takeover of industry and in formulating German demands. The state-owned Reichswerke 'Hermann Göring' became the instrument for running the captured industry. In Austria, Czechoslovakia, Poland and then in Belgium and eastern France, the Reichswerke absorbed the iron, steel, coal and machinery firms, until it became the world's largest industrial corporation, valued at RM 4,375 million in 1944. The most important acquisition was the famous Czech Skoda armaments works, in which the Reichswerke had bought a 63% stake by the end of 1940.

In the rest of occupied Europe the local economy was mainly exploited to meet the needs of the occupying forces, and to finance the German military effort. After the war it was calculated that the occupied areas contributed some 72 billion marks, 53 billion in occupation costs. The whole of German-

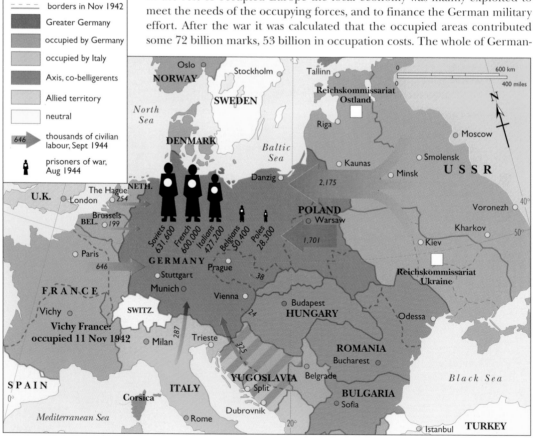

I/Foreign labour in Germany, 1944

- – – – borders in Nov 1942
- Greater Germany
- occupied by Germany
- occupied by Italy
- Axis, co-belligerents
- Allied territory
- neutral
- *646* thousands of civilian labour, Sept 1944
- prisoners of war, Aug 1944

Grammes of bread per week for a normal consumer, 1942

Denmark	2,280
Bulgaria	2,100
Germany	2,000
Czech Protectorate	2,000
Hungary	2,000
France	1,925
Serbia	1,800
Netherlands	1,800
Finland	1,750
Belgium	1,570
Lithuania	1,750
Latvia	1,700
Slovakia	1,670
Norway	1,645
Romania	1,500
Poland	1,490
Croatia	1,400
Italy	1,050
Greece	1,000

Meat consumption in grammes per week, normal consumer, 1942

Germany	300
Italy	100-200
Belgium	245
Bulgaria	unrationed
Czech Protectorate	300
Slovakia	300
Denmark	unrationed
Finland	70
France	180
Hungary	unrationed
Latvia	250
Lithuania	300
Netherlands	225
Norway	unrationed
Poland	130
Romania	250
Croatia	300
Serbia	200

European financial contributions to the German war effort
in millions of Reichmarks

France	34,200
Italy	13,300
Netherlands	10,078
Belgium	5,840
Norway	2,940
Poland	2,175
Bohemia	2,020
Occupied East	1,100
Serbia	472
Greece	95

Percentage of pre-war food consumption rationed by 1942

Germany	97
Czechslovakia	97
Poland	97
Belgium	97
Netherlands	96
Norway	95
France	95
Italy	95
Finland	90

controlled Europe, including satellite states, contributed 120 billion marks, or approximately 20% of the money spent on the war. As the war went on these areas also began to supply more manufactured products, either consumer goods which were no longer produced in Germany, or, from 1943, armament products as well. The French aircraft industry supplied 2,517 aircraft, Czechoslovakia 4,147.

The decentralization of production to the occupied areas encountered numerous problems. In many cases machinery and raw material stocks had been seized by the initial occupation authority, and new machines had to be supplied by Germany. The chief shortage was of labour, much of which was sent during the war to work in Germany. At the peak in 1944 there were 7.6 million foreign workers in Germany, 5.7 million civilians and 1.9 million prisoners-of-war. They came from all over Europe (some as voluntary workers seeking high wages in German factories), but the bulk were from the USSR (36%) and Poland (28%), forced, often at the point of a gun, to take work in German heavy industry, mining and agriculture. By August 1944 46% of the working population on the land in Germany was made up of foreign workers and prisoners, and one-third of the mining workforce. Although most were paid a low wage they lived in poor conditions in camps and barracks close to where they worked, on a meagre diet. Contact with the German population was forbidden, and sexual encounters were punished by death. The Gestapo made examples of foreign workers found guilty of slacking or insubordination. The German authorities found they could coerce and terrorize the foreign workforce more easily than the German, and the highest productivity gains were made in German industry at the point where the proportion of foreign workers was at its highest, in 1943 and 1944.

2/The Reichswerke

■ existing plants taken over by Reichswerke

products

C	coal
F	iron
S	steel
E	engineering
A	armaments
O	other

Resistance and Terror in Europe

Outside the Reich there was more opportunity to fight back against the German occupiers. Partisans and resistance fighters fought a 'shadow war' across Europe and provoked savage reprisals.

"To prevent a revolution…kill off the whole anti-social rabble."
Adolf Hitler, July 1942

After 1939 there were greater opportunities for resistance in the conquered areas than in the Reich itself. Resistance took a variety of forms, from minor acts of defiance or dissent to the mobilization of partisan armies. The German reaction was uniformly brutal, but it was particularly barbarous in the east.

Opportunities for non-violent resistance were limited but often more effective than armed resistance which invited a vicious reaction. In the Netherlands, for instance, three major strikes were mounted in 1941, 1943 and on the railways in 1944, to support the Allied invasion. Railway resistance was widespread in France. Public protests against the seizure of Jewish populations achieved little except in Bulgaria, which was not an occupied state.

Armed resistance was harder to organize and more dangerous. In France sporadic terror attacks were mounted by what became the *Armée Secrète*. It recruited from political dissidents and patriots, and from 1943 increasingly from young French workers fleeing the round-up of forced labour. The largest armed resistance came in Yugoslavia and the Soviet Union. In the Soviet Union partisan activity was eventually co-ordinated by the NKVD which tried to combine partisan sabotage and raids with the main battles at the front. German troops were ordered to show no mercy toward partisans, either in the USSR or elsewhere, and a campaign of mass killings, village burnings and hostage-taking was organized which decimated partisan ranks and alienated the native population from them.

A woman is shot at point-blank range, one of thousands subject to summary justice by German forces of occupation.

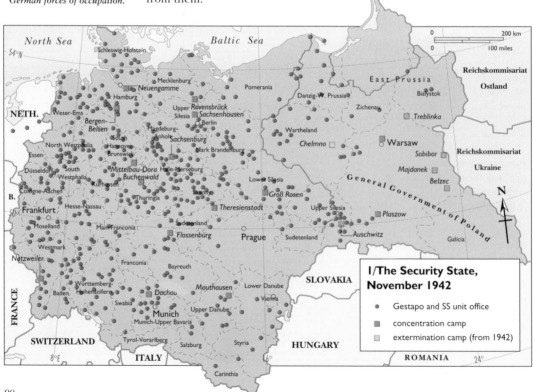

I/The Security State, November 1942

- Gestapo and SS unit office
- ■ concentration camp
- ■ extermination camp (from 1942)

3/Yugoslavia, mid-1943

- under partisan control
- under Axis attack

In Yugoslavia the inhospitable mountainous terrain gave opportunity for a sustained guerilla campaign, but divisions between the main communist-led guerilla movement under Tito and the nationalist Serb movement seriously weakened the response to occupation. German counter-measures proved successful in 1941 and 1942, and in 1944 Tito was almost captured in a daring raid on his headquarters. Only imminent defeat altered the prospects for successful resistance, as local populations abandoned collaboration to side with the likely victors. In Paris the resistance staged an uprising days before the arrival of Allied forces. In Yugoslavia the nationalities pooled their differences and in 1945 a national army drove the Germans out of most of the country.

In Poland two risings were staged, both in Warsaw. The first was a rising of the Jewish population in the ghetto. On 19 April 1943 the 60,000 remaining Jews in the ghetto fought a do-or-die battle against 3,000 troops. The rising was brutally suppressed. Approximately 50 Jews escaped. The second rising occured on 1 August 1944 as the Red Army approached Warsaw. The Polish Home Army mounted a rising with some 37,000 insurgents. The battle lasted 63 days, and ended with a complete German victory. About 250,000 Poles were killed, and in retaliation German forces destroyed some 83% of the city. The Red Army stood and watched.

2/Resistance to the Reich

- maximum extent of Axis expansion
- Greater Germany
- occupied by Germany
- occupied by Italy
- Axis co-belligerents
- Allied territory
- neutral
- ■ German civil administration
- ▲ German military administration

- ● security bases and detention centres
- ✳ major resistance incident in western Europe
- Polish resistance area
- Soviet resistance area
- ✡ armed uprising in Jewish ghettos
- major reprisal against civilian population
- escapees:
  - ● collecting centre
  - escape route

Genocide

With the introduction in 1941 of state-organized genocide the war gave the opportunity to complete the programme of 'racial cleansing' begun during the 1930s.

"The Jews shall be annihilated in our land."
Hitler to
Chvalkovsky, Jan
1940

The conquest of continental Europe provided the circumstances for a sharp change in direction in German race policy away from discrimination and terror to the active pursuit of genocide. Hitler and the racist radicals in the Nazi movement had no master plan for annihilation in 1939, but their whole conception of the war was one of racial struggle in which the Jewish people above all were the enemy of German imperialism. When Germany found herself ruling very large Jewish populations after the conquest of the east the regime began to explore more extreme solutions to the 'Jewish question'.

The German New Order was viewed from Berlin in terms of a hierarchy of races: at the apex were the Germanic peoples, followed by subordinate Latin and slavic populations, and at the foot were races—Jews, Sinti and Roma (gypsies)—who were deemed to be unworthy of existence. The treatment of these racial groups began with a programme of ghetto-building or imprisonment, but in the summer of 1941 became suddenly more violent. The orders prepared for *Barbarossa* deliberately encouraged the murder of Soviet Jews. In the Baltic states and Ukraine native anti-semitism was whipped up by the German occupiers and led to widespread massacres. In July 1941 there is strong evidence that Hitler at last ordered the 'physical extermination of the Jews' (Adolf Eichmann), flushed with the prospect of victory in the USSR

1/Sinti and Roma deaths, 1939–1945

Axis/pro-Axis territory	20,000 Sinti and Roma populations in 1939
Allied territory	
neutral	1,000 numbers murdered

Jews are massacred in Kaunas, Lithuania c. summer 1941. Before the first extermination camps were set up Jews were rounded up and murdered in the areas where they lived.

and the realisation that there were no longer any forces in Europe which could constrain a programme of annihilation.

Extermination was placed on a proper organizational foundation with the establishment under the RSHA (*Reichssicherheitshauptamt*) of a series of camps in occupied Poland where victims were either gassed as soon as they arrived, or worked to death in factories and quarries close to the camps. The systematic murder of Jews began in late 1941, and was extended to the Sinti and Roma in 1942. In 1943 and 1944 Germany put pressure on Italy and Hungary, her allies, to release their Jewish populations. When both states were occupied by German forces any resistance to anti-semitism was quashed and hundreds and thousands of Jews were slaughtered in 1944 and 1945 in an effort to complete what Hitler saw as his chief legacy for Europe, a 'Jew-free' continent. Records remain inadequate, but approximately 5.3 million Jews and at least 250,000 Sinti and Roma were killed.

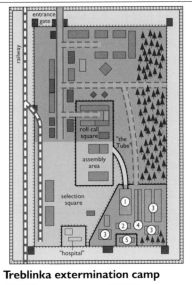

Treblinka extermination camp

- staff areas and quarters
- auxiliary staff area
- reception camp
- extermination sector
- storerooms and service buildings
- staff barracks
- workrooms
- processing barracks
- latrines
- machine gun towers
- barbed wire fence
- ① new gas chambers (10)
- ② old gas chambers (3)
- ③ mass graves
- ④ area for burning bodies
- ⑤ area for burning bodies

2/Jewish deaths, 1939–1945

- area of gas van operations
- sites of euthanasia operations
- extermination camps from 1942
- 1,000 estimated Jewish deaths
- main rail links to extermination camps

FINLAND 15

ESTONIA 1,000

728 NORWAY

SWEDEN

DENMARK 77

LATVIA 80,000

LITHUANIA 143,000

DANZIG 1,000

UNITED KINGDOM

NETH. 106,000

POLAND 3,060,000

Treblinka

U S S R 1,200,000

BELGIUM 24,387

LUX. 700

GERMANY 160,000

Chelmno

Sobibor

CZECHOSLOVAKIA 217,000

Majdanek

Belzec

Auschwitz/Birkenau

AUSTRIA 65,000

FRANCE 83,000

SWITZERLAND

HUNGARY 200,000

ROMANIA 269,632

Black Sea

ITALY 8,000

YUGOSLAVIA 67,122

BULGARIA

ALBANIA 200

TURKEY

GREECE 71,301

FRENCH NORTH AFRICA

Mediterranean Sea

LIBYA 582

Cyprus

North Sea

Baltic Sea

0 800 km

0 500 miles

VI: German Society and Total War

German society was mobilized to fight a total war by a Party which turned the terror on its people as defeat loomed.

"Do you want total war? ...Do you agree that anybody who injures our war effort should be put to death?"
Joseph Goebbels,
February 1943

The war had profound affects on German society and economic life. The home economy was closely controlled and a comprehensive system of rationing imposed. Thirteen million men were taken out of the workforce, half of them from agriculture and commerce. Women were forced to take on many of the tasks they had shared with men, particularly on the land. Popular racism surfaced, partly in response to the crude racist ideology peddled by the regime, partly to the difficult circumstances created by the influx of more than seven million foreign workers. Finally, the bombing came to dominate the lives of German city-dwellers. Millions left for the villages and small towns. The welfare services of the regime were stretched to the limit coping with evacuation and rehabilitation.

The extensive mobilization of the regime was announced in the War Economy Decree published on 4 September 1939. Under its terms rationing was imposed on the civilian economy across a whole range of products. The production of anything but the most essential consumer goods was closed down. A limited and largely carbohydrate diet was made available for all. Meat, fats and fresh food were cut back, and supplies diverted for the use of the armed forces. The average consumer had to make do with less than half a pound of meat a week, and one egg. Even the supply of these goods depended on availability. From 1939 most German civilians enjoyed a monotonous diet of black bread, potatoes and *ersatz* or substitute foods. The regime was determined to avoid a repeat of the price inflation and black market of the First World War. Black marketeering was heavily punished, in some cases with execution. Ration cards were available for everyone, rich or poor, on exactly the same basis. Peasants could consume their own food, or favour those they knew, but the state policed food production much more closely and effectively than in the previous conflict. There was no starvation in Germany during the Second World War, though the ration had to be cut at times. The provision of what was called an 'existence minimum' prevented serious social tensions.

A visiting American in Berlin in 1940 described conditions as 'Spartan throughout'. In cities hot water was rationed to a couple of days a week; train travel was difficult to arrange and lengthy; families were permitted a bar of *ersatz* soap a month; by 1941 women were rationed to one and a half cigarettes a day, men to three. These conditions remained in the main constant to 1944 when the bombing made the distribution of goods more difficult and destroyed a great many of the shops at which consumers were registered for rations. Emergency services supplied water, food and clothing for bombed-out families and undertook to supply a small quantity of replacement household goods, which necessitated a slight expansion of civilian production in 1943 and 1944. In the last year of war the economy was facing widespread difficulties due to the bombing. Production was concentrated on essential weapons and civilians took a back seat. By the end even the 'existence minimum' could not be supplied and by the time Allied troops arrived hunger had become a serious problem. Civilian shortages help to explain why foreign workers and camp prisoners died in such numbers. They were at the end of the German food chain, and supplies destined for them were stolen or sold.

A poster warning Germans to observe the blackout: 'The Enemy Sees Your Light'. Extensive civil defence preparations were made during the 1930s, with a large programme of air-raid construction and training in civil defence measures. Casualties would almost certainly have been much higher in the bombed cities without them.

A painting by Frank Wooton of the RAF Bomber Command raid on the Möhne Dam on the night of 16/17 May 1943. The precision attack successfully breached the dam and caused widespread loss of life and destruction. The attack was not repeated and within the month the damage had been repaired.

The changing conditions of social life affected German women more than men. There remains a persistent myth that German women stayed home during the war while their American and British sisters worked in munitions factories. Women played a large and essential part in the German war effort. They already formed a substantial share of the workforce before 1939, when the rearmament boom brought many more into factory employment. By 1939 the female share of the workforce was already higher in Germany than it reached in Britain or the USA during the years of wartime employment. All single women were required to work, or to perform Labour Service, or to contribute to volunteer work. Six million women worked in German agriculture, which because of its small scale farms and low level of mechanization required a large labour input. The withdrawal of men from the villages—peasants supplied the single largest source of military recruits—left women and youths to cope until foreign workers and POWs were sent to the countryside from 1941. In small businesses such as shops or hotels it was not uncommon for the women to take on the running of the firm. Millions of women were also used to take the place of men in transport, administration, communications and commerce. Since so many were already in employment in 1939 this involved a substantial redeployment of women, a move obscured by the gross figures on female employment. By 1944 three and a half million women worked part-time in addition to the 14.8 million already working.

It was inevitable in a country where 13 million males were withdrawn into military activities that women would have to take a larger part. The Party itself became a recruiting agent for popular participation in the war effort. During the war the number of Party members leaped up to more than eight million. Many new recruits came from working class backgrounds, and larger numbers of women also joined. The SA and HJ were used conspicuously to help with the bomb damage and with rehabilitation, or to organize collections of scrap metals and material. The Party Gau organization began to assume a greater political significance as the number of state officials declined and the organization of local activities devolved onto party offices. In industry the Labour Front interfered extensively in relations between management and labour, and by no means in favour of the former. Working-class recruitment into the Party owed something to a shift in favour of work-

The ruins of Berlin in 1945 with the Brandenburg Gate in the background. The city was hit by 24 major bomb attacks which destroyed one-third of all housing and reduced the population from 4.3 to 2.6 million as the bombed-out families moved to safer parts of the country. The Soviet offensive in April 1945 reduced large stretches of central Berlin (where it was still standing) to rubble.

ers in the industrial balance of power, a result, as in the First World War, of labour scarcity. This process also owed something to the realization by workers that the party could be used as an ally to put pressure on employers, who were not much liked by the Party bureaucracy. The bargaining power of labour cannot be established with any certainty because of the nature of the autocratic system which denied labour political or trade representation. It was a power that almost certainly declined towards the end of the war with the influx of foreign workers who accepted lower wages and poorer conditions, and who could be made to match the mass production methods being introduced into arms factories. Both the German and foreign workforces found themselves subject to a more arbitrary and coercive structure of authority as the security services came to play a growing part in mobilizing and sustaining the last production efforts in 1944 and 1945.

Popular attitudes to the regime rose and fell with the success of German arms. Secret-police monitoring observed a slackening of enthusiasm from 1941 with the onset of war with Russia, which was not popularly welcomed like the war with Poland. The turning point came with the Stalingrad defeat in January 1943 and the destruction of Hamburg in July 1943 when confidence in victory evaporated. There was an increase in covert resistance activity, but public patriotism, fear of the authorities, and the mounting brutality of the security services made resistance both dangerous and ineffective. The 'White Rose' movement in Munich exemplified the problem. It was organized in the university by Hans and Sophie Scholl, among others, in 1942. They mounted a street demonstration in Munich in February 1943. Betrayed to the Gestapo, they were both executed along with fellow resistance fighters on 22 February 1943. A major communist spy-ring centred on Göring's Air Ministry, and known as the Red Orchestra (Rote Kapelle) was broken up in 1942 and its 100 or so members imprisoned, tortured and exe-

Women were awarded the bronze Mother's Cross for bearing five children for the Fatherland. Women were encouraged to have large families to provide soldiers and colonists for the new German empire.

cuted. For those who contemplated even minor acts of dissent—listening to outlawed foreign broadcasts, for example—the punishments were draconian. The Gestapo, which was under-manned for the tasks allotted to it during the war, resorted to summary imprisonment without trial even for trivial offences, as it struggled to cope with the paperwork of formal judicial proceedings. When in 1942 the Justice Ministry was taken over by Otto Thierack, a Himmler appointee, the entire legal system was turned into the arbitrary agent of the most radical element of the movement.

The final year of war saw state lawlessness institutionalized. The People's Court set up in Berlin to try cases of political resistance, presided over by Roland Freisler, sat *in camera*, while the prosecutors bullied prisoners into confessing political crimes, like the Stalinist show trials of the 1930s. In February 1945 the courtroom was demolished in a bomb attack and Freisler killed. In the last weeks of war the SS and Party extremists took a final revenge on prisoners and dissidents. Thousands were murdered as Allied armies approached. Thousands more died in the final bomb attacks against almost undefended cities, crammed with refugees and evacuees. Ordinary Germans became obsessed with sheer survival. There was no 'stab in the back' from the home front, which Hitler had always used to explain defeat in 1918. Soldiers and civilians alike became the victims of a final orgy of terror from a Party machine which had traded on intimidation and violence from its inception.

Field Marshal Erhard Milch talks to a woman aircraft factory worker on a tour of inspection. By 1944 2.8 million women worked in German industry, 1.5 million of them in heavy industry and engineering vital to the war effort.

Mobilizing for Total War

During the war the regime mobilized the economic and labour resources of the state community to the full, a process that began in 1939.

"The total mobilization of the economy has been ordered..."
Colonel Georg Thomas, 3 September,1939

When war broke out with Britain and France Hitler ordered the full mobilization of the German economy. Throughout the 1930s both Hitler and the armed forces had argued that a future war could only be won by making the greatest use of the material and human resources available whether the conflict was a long or a short one.

Responsibility for transforming the German economy was divided between a number of agencies, none of which was able to impose any kind of central direction. The armed forces saw the militarized economy as their affair, and through a system of armaments inspectorates both in Germany and in the conquered areas, they tried to organize the production of armaments and

1/Organization of Economics and Armaments Office, OKW, 1942

- German Reich
- protectorate and general government
- German occupied

IX Kassel German inspectorate

General government foreign inspectorate

2/Men in reserved occupation workforce, 1942

(as a percentage of all employed men)

- 45
- 42
- 41
- 38

The Nazi military hierarchy insisted on a direct involvement with the armaments industry.

the diversion of resources. They were in direct competition with Göring's Four Year Plan organization, the Economics and Labour ministries, and the local party *Gaue*, all of which had some claim on economic policy. In March 1940 Hitler oppointed Fritz Todt as Armaments Minister, but his responsibility extended only to army weapons. He became just another element in what General Thomas, head of the OKW War Economy and Armaments Bureau, called 'the war of all against all'.

The framework for a war economy was established from the start of the war. All major consumer goods were rationed from September to ensure that the population all got a fair share of food and basic goods. Thousands of civilian businesses were shut down or converted to war production. Taxes were raised, doubling tax revenue in four years. By the summer of 1941 some 6.6 million men had been taken out of the economy into the armed forces, leaving women to run Germany's numerous small peasant farms and small businesses, or to transfer from producing consumer goods to working on war orders. By 1941 almost 70% of those working in manufacturing industry were producing goods for the armed forces. In the consumer goods sectors (except food) between 40 and 50% of production went to the military, leaving extremely limited quantities for the civilian population. By the summer of 1941 Thomas reported that Germany was close to full economic mobilization.

Paradoxically the transfer of resources to the military economy did not produce a corresponding increase in the output of weapons. The confused organization and military demands for excessively high quality produced a very inefficient war economy. In 1941 Hitler ordered the thorough rationalization of the system in order to extract more goods from the same resources. In February 1942 the architect Albert Speer was appointed by Hitler to be Minister for Armaments and Munitions with orders to complete the programme of rationalization. He centralized the organization of the war economy as far as he could, and he pushed through a programme of rationalization which trebled armaments output in three years with a relatively small increase in resources, and with the loss of millions more skilled workers, a loss that would have been greater still without the introduction of a system of reserved occupations which gave the most important skilled men exemption from military service. By 1944 women made up 51% of the native German workforce, and the civilian economy was cut to the minimum. It was at this point that Speer's efforts were finally frustrated by large-scale bombing.

Military expenditure

in billion Reich Marks

80	
60	
40	
20	
0	

1938 39 40 41 42 43

3/Conscripts from industry, 1942

(as a percentage of 1939 male labour force)

- 33
- 30
- 26

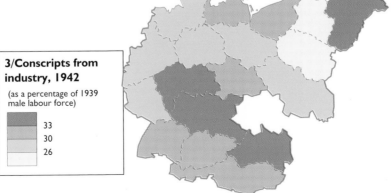

The Rise of the SS

The war gave the SS an opportunity to increase its influence in every area of life, both in Germany and in the occupied states of Europe.

"I know there are many people in Germany who feel ill if they see this black uniform; we understand this and do not expect to be loved."

Heinrich Himmler, 1936

The SS (*Schutzstaffel*, or security squad) began life as Hitler's personal body-guard, chosen from especially loyal party members. In 1929, when it numbered only 280 members, it was taken over by the young Heinrich Himmler. He turned the organization into the élite of the movement, in its distinctive black uniform and death's head insignia. By 1939 it numbered some 240,000 men, organized in divisions and regiments. Its sphere of activity embraced the Reich security and concentration camp system, a range of economic activities (including the production of mineral water and jam) and, from November 1939, an Armed (*Waffen*)-SS which provided some 38 divisions and 800,000 men by the end of the war, fighting side by side with the regular army.

During the war the SS increased its responsibilities in all these areas. The number of concentration camps for political prisoners expanded enormous-ly. The camp population of 25,000 reached 700,000 by the end of the war, but thousands of others died of disease and hunger. The SS also ran the exter-mination camps set up in the east for the mass-murder of Jews and gypsies.

I/The SS in Europe

- Greater Germany
- occupied by Germany
- occupied by Italy
- Axis co-belligerents
- Allied territory
- neutral
- —— border in Nov 1942

executive branch of security police and SD

security police and SD command

2/Lebensborn organization

■ central office
● branch office
▲ maternity home

The huge prisoner population provided Himmler with the instrument to increase SS activities in the economy. His camp inmates were used for building and agricultural labour, and, as the war went on, for armaments projects as well. In 1943 Himmler promised to produce the V-weapons for Hitler with his camp labour and to build the underground factories and installations which Hitler wanted to avoid the bombing. Both projects took a vast workforce and pushed the SS into the centre of the war economy. In 1944 the Speer Ministry was gradually eased out of running the industrial effort and Oswald Pohl and Hans Kammler, the leading SS economic authorities, took over the emergency production programmes, which needed forced labour and harsh methods to sustain them.

The SS played the main part in organizing the Final Solution of the Jewish question, and in running the apparatus of racial hygiene. This included the Lebensborn organization which was set up to provide healthcare and welfare for German women in one of twelve clinics if they chose to bear the child of a 'pure' SS man. During the war the organization became involved in seizing children with 'aryan' features in the occupied east in order to improve the German bloodstock. An estimated 300,000 children were sent to Lebensborn centres from where they went on to adoption by German families.

The SS became in the last year of the war, the driving force of the regime, dealing out a summary justice, fighting with a fatalistic savagery at the front and moving its army of wretched slave labourers from project to project as the Allies closed in on the Reich. When Himmler was captured by British soldiers at the end of the war he committed suicide.

Reinhard Heydrich (1904–42), seen here on Himmler's left, was his right hand man and head of the RSHA.

3/The SS during the war

—— SS regions (*Wehrkreise*)

🔲 sites of divisions 81-125 of general SS

SS enterprises

🏠 construction material industry

🥫 food industry

🪵 wood working industry

■ other

▪ concentration camp

A armaments production

G gravel and stone-working

The Impact of Bombing

In the last two years of war bombing became the chief factor affecting the urban population of Germany, creating widespread dislocation and social hardship.

"The air war has now turned into a crazy orgy ... the Reich will be turned into a complete desert."
Joseph Goebbels, 1945

In the last two years of the war the German home front was dominated by the effects of bombing. War production was affected by direct bomb attack and by the diversion of resources. Urban society was transformed by the large-scale evacuation of cities, the massive destruction of housing and amenities, and the almost constant state of alarm. Bombing strained emergency services and the German welfare system almost to breaking point. Post-war surveys of German morale showed that 91% of Germans believed that bombing was the hardest thing for civilians to endure in the war.

War production was inhibited by bombing in a number of ways. The direct attack on particular industries (oil, chemicals, aviation, vehicles), became significant in 1944 and 1945. German synthetic oil production was reduced to only 5% of output by September 1944; the output of chemicals for explosives was reduced by three-quarters in 1944; the output of synthetic rubber by 88%. The production of finished weapons was kept going through stringent rationalization of production methods and the dispersal of contracts to areas safer from bombing, but here too far less was produced than was planned: around one-third of the output of tanks, lorries and aircraft was lost in 1944.

The indirect effects were as severe. Some two million workers were engaged in anti-aircraft defence, in clearing up bombed cities and factories and the rehabilitation programmes. The dispersal programme in 1944 was based on a vast system of underground factories which took up half a million

Results of bombing major cities

Killed:	305,000
Wounded:	780,000
Homes destroyed:	1,865,000
Persons evacuated:	4,885,000
Person deprived of utilities:	20,000,000

28% of all homes damaged or destroyed.

22 million Germans subjected to bomb attack.

Towns attacked by size of attack and effects on population

Weight of bombs (tons)	Number of cities	Population (millions) 1939	1945
20,000+	15	9.0	4.9
10–20,000	11	1.7	1.1
5–10,000	19	2.1	1.5
1–5,000	53	2.8	2.1
100–1,000	54	1.5	1.4

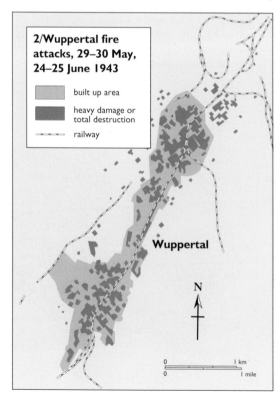

2/Wuppertal fire attacks, 29–30 May, 24–25 June 1943

built up area

heavy damage or total destruction

railway

Wuppertal

N

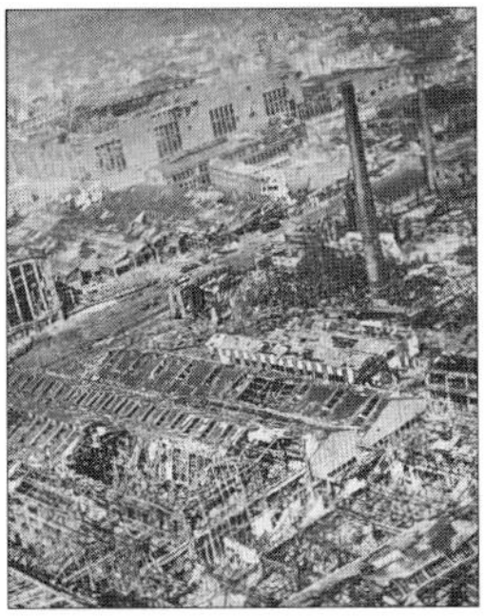

The Krupp works at the end of the war. This view of Essen shows the extensive damage. The city was hit by 28 major raids, more than any other. At the heart of Ruhr industry, Essen was a major target of British Bomber Command from 1940. Much of its production was moved to other parts of the Reich. 100,000 dwellings were destroyed.

construction workers, and wasted a great deal of equipment and resources for very little return. The fight against the bombers absorbed one-third of all gun production, and one-third or more of radar, optical and electro-technical output. The direct and indirect effects on war production reduced the potential output of weapons for the battlefields by approximately 50%.

The social consequences of bomb attack also reduced economic performance. Workers in cities spent long hours huddled in air-raid shelters; they arrived for work tired and nervous. The average rate of absenteeism in 1944 among German workers was 23% on any one day. Alarms sounded day and night as the bomber streams flew overhead. In Mainz alarms were in force for a total of 540 hours in 1944, the equivalent of five or six weeks work. The effects of bombing in the cities also reduced the prospects of increasing female labour as women worked to salvage wrecked homes, or took charge of evacuated children, or simply left for the countryside where conditions were safer. In the villages there developed serious tension between the peasantry and the influx of townspeople, over issues such as food or crime. The flood of refugees from bombing strained the rationing system, while hospitals had to cope with three-quarters of a million casualties. Under these circumstances demoralization was widespread, though the terror state and the sheer struggle to survive prevented any prospect of serious domestic unrest.

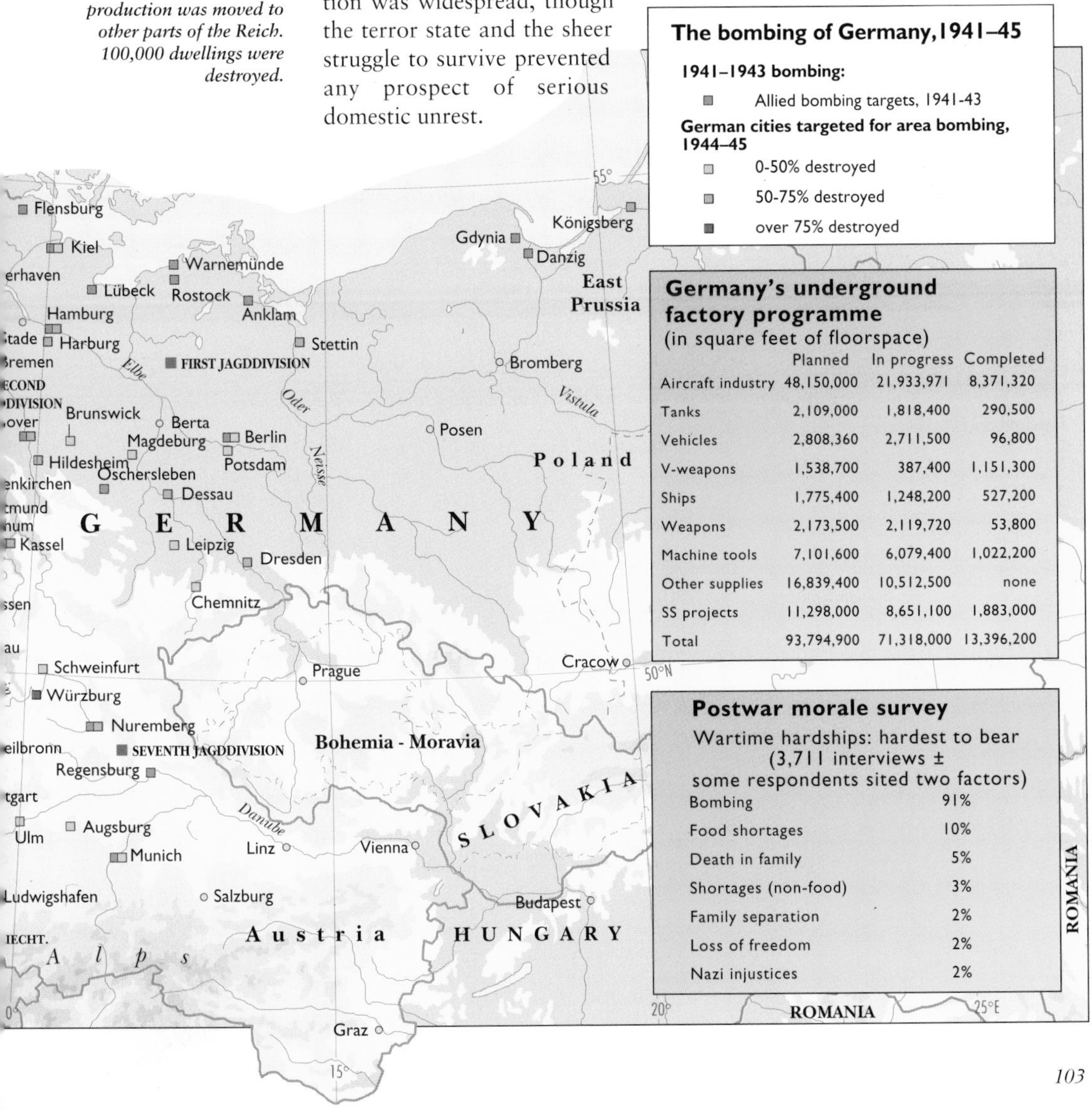
The bombing of Germany, 1941–45
1941–1943 bombing:
Allied bombing targets, 1941-43
German cities targeted for area bombing, 1944–45
0-50% destroyed
50-75% destroyed
over 75% destroyed
Flensburg
Kiel
Warnemünde
Lübeck
Rostock
Anklam
Hamburg
Harburg
Stettin
FIRST JAGDDIVISION
Elbe
Oder
Brunswick
Berta
Magdeburg
Berlin
Hildesheim
Oschersleben
Potsdam
Dessau
Neisse
G E R M A N Y
Kassel
Leipzig
Dresden
Chemnitz
Schweinfurt
Würzburg
Nuremberg
SEVENTH JAGDDIVISION
Regensburg
Augsburg
Ulm
Munich
Ludwigshafen
Salzburg
Linz
Danube
A l p s
Austria
Graz
Gdynia
Danzig
Königsberg
East Prussia
Bromberg
Vistula
Posen
Poland
Prague
Bohemia - Moravia
Cracow
SLOVAKIA
Vienna
Budapest
HUNGARY
ROMANIA
55°
50°N
20°
25°E
15°

Germany's underground factory programme
(in square feet of floorspace)

	Planned	In progress	Completed
Aircraft industry	48,150,000	21,933,971	8,371,320
Tanks	2,109,000	1,818,400	290,500
Vehicles	2,808,360	2,711,500	96,800
V-weapons	1,538,700	387,400	1,151,300
Ships	1,775,400	1,248,200	527,200
Weapons	2,173,500	2,119,720	53,800
Machine tools	7,101,600	6,079,400	1,022,200
Other supplies	16,839,400	10,512,500	none
SS projects	11,298,000	8,651,100	1,883,000
Total	93,794,900	71,318,000	13,396,200

Postwar morale survey
Wartime hardships: hardest to bear (3,711 interviews ± some respondents sited two factors)

Bombing	91%
Food shortages	10%
Death in family	5%
Shortages (non-food)	3%
Family separation	2%
Loss of freedom	2%
Nazi injustices	2%

Resistance, Terror and Collapse

As the war drew to a close the Nazi regime imposed even more brutal terror on the German people, provoking a generals' revolt in July 1944.

"I myself set off the bomb during the conference with Hitler...No one who was in that room can still be alive."
Col. Claus von Stauffenberg, July 20, 1944

A post-war survey of popular opinion in Germany during the later stages of the war showed that by January 1944 77% of the sample regarded the war as lost. Secret police reports began to reveal to the German authorities sharply declining enthusiasm for the war. The regime's response was to turn the screw of repression tighter still. In the last stages of the war many more native Germans, including men as distinguished as the banker Hjalmar Schacht, master-mind of the economic revival, were taken into custody. The concentration camp system had approximately 25,000 prisoners in 1939; in 1945 the number was more than 700,000, including workers sent to 'education' centres for alleged saboteurs and slackers. As the ring tightened around Germany, so the SS closed down the outlying camps and drove their prisoners on long death marches to the interior.

Under the pressure of imminent defeat new areas of resistance began to crystallize. The most significant hostility came from the military and gentry elite, who resented their increasing exclusion from social influence and political power and saw themselves as the natural defenders of German national interest, which Hitler's reckless war-making threatened to destroy. Some were genuine enthusiasts of democracy or even socialism; others were hoping for the restoration of a conservative Germany untainted by the vulgarity and racism of the Nazis. The 'Kreisau Circle' around Count Helmut von Moltke

Camps 1939–45

- main camps
- smaller camps/ Gestapo prisons
- worker education camps
- foreign worker camps
- courts

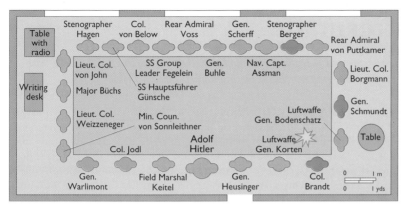

Göring and Bormann examine the damage done to Hitler's conference room after the bomb blast of 20 July 1944.

2/1944 Bomb Plot

The Führer's map room (above)

killed

injured

position of bomb

Stauffenberg's route (right)

on foot

by car

mined area

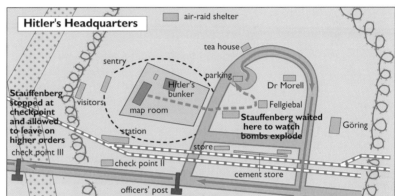

was broken in January 1944 by the Gestapo (and Moltke himself arrested and later executed), but enough conspirators survived to mount a serious assassination attempt in July 1944. A German cavalry officer, Count Claus von Stauffenberg, set out on 20 July for Hitler's headquarters in East Prussia. He planted a bomb in a briefcase beneath the heavy oak map table over which Hitler was poring, went outside to watch the blast from a fellow conspirator's office, and then hurried from the complex back to Berlin. The plan was to set up an alternative government and to try to sue for peace with the West. Hitler survived the blast (the briefcase was pushed aside at the last moment), and neither the rest of the army nor the country was prepared to follow the conspirators' lead.

In the wake of the assassination attempt the Gestapo rounded up all known suspects and the Reich descended into an orgy of lawlessness and killing as fanatical Nazis took their revenge on any individuals or groups regarded as defeatist or anti-Nazi. Production was kept going for the war effort by extracting the last ounce of work-power from hundreds of thousands of slave labourers in the camps, and millions of foreign workers whose conditions came to resemble more closely those of the camps. The Reich fragmented into a number of self-contained economic areas as the bombing destroyed rail and water transport. Factories lived off accumulated stocks. By March 1945 the economy was close to collapse. Hitler's closest companions began to search for an avenue of escape. Himmler set up contacts to the Allied powers; Göring tried to usurp Hitler's authority; Speer, his Armaments Minister, countermanded Hitler's orders for a scorched earth policy across Germany as German forces retreated to a last-ditch stand around the capital. When Hitler killed himself on 30 April in Berlin, his Reich was in ruins.

VII: The Aftermath

After the war Germany was partitioned along the lines of the Cold War conflict and was not finally reunited until 1990.

> *"In the rebuilding of Germany the great task consisted in awakening and strengthening the democratic forces in the population and in allowing them to grow"*
> Konrad Adenauer,
> *Memoirs*

The complete defeat of Hitler's Germany in May 1945 left a Europe quite literally in ruins. In the Soviet Union 70,000 villages and 1,700 cities were destroyed. France lost an estimated 40% of her national wealth, Italy one-third. In Germany the combination of bomb damage and the Allies' artillery turned much of Germany's urban area into a bleak moonscape of craters and fractured buildings. By 1944 some 62 billion RM of losses in buildings and equipment had been reported. In the major cities of the Reich an average of between 55 and 60% of all dwellings were destroyed. For years after 1945 Germans lived in the cellars and surviving shelter of the ruined streets, prey to high levels of disease and crime.

The victorious Allies did not behave as they had done in 1918. This time the whole German and Austrian area was occupied by Allied military and civilian personnel who took initial control of all areas of German life. Many Germans had been subjected in the later stages of the war to propaganda that told them to expect the very worst from the occupying powers. The experience in the east, where the Red Army exacted a very real physical revenge on the civilian population, lent weight to this fear. The western Allies were divided on quite how severe to be with Germany, but Roosevelt's Secretary of the Treasury, Henry Morgenthau, favoured a policy of enforced ruralization,

German Military Government Police armed with wooden truncheons and under British command, Dannenberg, July 1945. Germans were employed for police and administrative duties by the occupying regimes, but were subject to a de-Nazification screening process.

permanently removing Germany's industrial base which had allowed her to wage war twice in a generation. Roosevelt and Churchill at first endorsed these harsh conditions, but were persuaded that a weak, economically backward Germany would not be in the long-term interests of either state. Nevertheless Germany was not to become a buoyant industrial economy again. In March 1946 a Level of Industry schedule was drawn up preventing Germany from producing a whole range of advanced industrial goods, and limiting steel production to 5.8 million tons (against 28 million in 1944). German disarmament went almost without saying.

The Allies wanted to avoid a long-term reparations bill like the one imposed in 1919, but they did seek reparation through the physical seizure of industrial and scientific assets and technological knowledge. The American rocket programme in the 1950s relied on the prior work done by the German rocket specialist Wernher von Braun, who was recruited to work in America. Britain took from Germany an exhaustive number of blueprints and inventions which were applied to British industry and helped Britain's economic revival after 1945. The USSR took machinery and factory equipment from its own zone of occupation in eastern Germany and from the zones of the other three occupying powers, Britain, the USA and France. The total value was estimated by the US authorities at more than 6 billion dollars between 1945 and 1948. The poor state of German agriculture following the collapse meant that a flow of resources passed from the Allies to Germany in the form of food and agricultural supplies. German requirements came at the foot of the list, however, and proved insufficient to meet the food needs of the population. The German people faced three years of food rations below the levels they had eaten during the war, and the more vulnerable members of the community faced severe hunger. There developed a lively black market in food. The official price of a loaf of bread in the British zone was 52 pfg; the black market price was 25 marks.

Left: A scene from the Nuremberg Trials of the major war criminals held between November 1945 and October 1946. Twenty two principal defendants were tried, and Martin Bormann, Hitler's secretary, was tried in absentia. Göring (seated, left) kept up a vigorous defence and admitted his support for Hitler, but not his guilt. He committed suicide on the night of his execution. Another nine were hanged, and Bormann was condemned to death but never caught.

The sheer cost to the Allies of sustaining the existence of the German population was one of the arguments used to justify a shift in attitude in the British and US Zones. In 1947 both authorities decided to encourage higher levels of German economic activity in order to get Germany to supply more of its

Trucks of the US Fifth Army return looted artworks to Florence in 1945. Millions of items of art, books and museum holdings were moved into store, or looted or destroyed during the war. Hitler and Göring were avid art collectors, and their agents scoured Europe looking for objects to buy at advantageous prices.

own needs. They looked to Germany as a source of future markets for western goods, but if it remained impoverished and repressed the west feared a revival of German communism in the western zones at a time when the political confrontation with the Communist Soviet Union was taking firm shape. For all these reasons the revival of the industrial economy in the western zones was deliberately stimulated. In 1948 the currency was established, and the western zones qualified for Marshall Aid under the European Recovery Programme. Between 1947 and 1952 the USA made $3.1 billion available in aid. By 1950 western Germany had recovered pre-war levels of output. By 1955 industrial output was double the level of 1936. A combination of favourable world market conditions and a highly motivated managerial élite and workforce created the economic miracle of the 1950s and 1960s. West Germany recovered the trajectory of development lost in 1914.

Germany's economic revival raised serious political issues. Soviet resentment at American and British economic policy led by 1949 to a complete partition of the German areas into a western and an eastern state, and ended any hope of a generally agreed peace settlement with Germany. In Eastern Germany a Stalinist dictatorship was imposed which bore strong resemblances to the pre-war system with its apparatus of secret policemen, censors and political hacks. For those Germans unfortunate enough to be caught in the crossfire of the Cold War there were another forty years of political repression to add to the dozen already experienced. Communist priorities were, however, different, economic development and social reform, though slower than in the

The Berlin Wall built to seal off West Berlin from the communist state that surrounded it was the most visible symbol of the legacy of partition after the Second World War. Built in 1961 by the East German regime of Walther Ulbricht it was finally removed in late 1989 when the communist system collapsed. On 3 October 1990 the two parts of Germany were formally re-unified under the Christian-Democrat government of Helmut Kohl.

western areas were real enough. Yet by the 1980s the gap between the two sides had grown unbridgeable, and the growing frustration of young East Germans with the economic inefficiency and political repression of the Communist state produced a revolutionary crisis in 1989 which resulted in the collapse of the communist regime in a matter of weeks. In October 1990 the two parts of Germany were re-united.

In the western part political development picked up where the crisis of liberal democracy had left it in the late 1920s, but with two important differences. The Basic Law of 1949 produced a constitution which allowed a Constitutional Court to ban parties committed to subverting parliamentary democracy, and removed the power of the President to rule by Emergency Decree. The system of small parties and weak coalition government in the 1920s was succeeded by a virtual two-party system run by Christian Democrats and Social Democrats both thoroughly committed to democracy, social welfare and the development of a 'social market economy' — capitalism with a caring face. Under the first post-war Chancellor, the veteran Catholic politician Konrad Adenauer, and his Economics Minister, Ludwig Erhard, the Christian-Democrat government concentrated on economic revival as the key to social stability. Internationally the new state built strong bridges to the liberal west. In 1955 the West German state joined NATO. Two years later Germany was a founder member of the European Economic Community. Relations with the United States, which had been poor in the 1930s, remained close. 'Americanization' was widespread in the 1950s as young Germans sought to put the Nazi past behind them.

The revival of a powerful German economy had none of the associations with imperialism and world power that it had had before 1914. None the less it was viewed by some of Germany's former enemies with alarm. The USSR made the frontier between the two Germanys the so-called Central Front of the Cold War, with West German cities in the firing line in any nuclear exchange. In the 1960s and 1970s ultra-nationalist movements reappeared in German politics, some with an overtly neo-Nazi complexion. Nothing like the *revanchisme* of the 1920s developed out of this revival; there is little evidence that the German population wanted an active foreign policy based on the recovery of the lost territories and a reassertion of German military power. There did emerge a sharp debate in the 1980s over the issue of coming to terms with Germany's Nazi past. In the *Historikerstreit* (conflict of histo-

German children watch an American transport plane land at the Tempelhof airfield in Berlin during the Berlin Air Lift from June 1948 to May 1949. Soviet efforts to drive the western powers out of Berlin were frustrated by a western aid programme which delivered 2.3 million tons of supplies in more than 200,000 flights.

rians) over interpretations of the Nazi period there emerged a significant polarization between those who wished to place the Third Reich in its historical context and end the long sense of collective German guilt, and those who saw the crimes of the Nazi regime as something which prevented them ever seeing the period as a 'normal one', which the German people could expunge from their conscience. The passions aroused by the arguments showed the extent to which the experience of the Third Reich has cast a long and malign shadow across the German present.

Occupation and State Formation

When Allied armies occupied Germany in 1945 there was little clear idea about what shape a future German state might take. By 1949 two states, one communist, one capitalist were in existence.

"It is not the intention of the Allies to destroy or enslave the German people."
Potsdam Conference declaration, August, 1945

With Germany's final defeat the Allied powers occupied Germany and Austria along lines they had agreed in discussions since 1943. A French occupation zone was included and the Potsdam Conference in July and August 1945 confirmed the territorial settlement. The frontiers of Poland were moved westwards to the Oder River, absorbing half of East Prussia and Silesia. The other half of East Prussia was ceded to the Soviet Union. The union of Austria and Germany was abrogated by a Soviet-backed government in Vienna on 29 April. Austria was treated as a 'liberated' state rather than as an integral part of the Third Reich. Democratic elections were held in November 1945. A full Treaty was not signed until 1955, when the last foreign troops on Austrian soil were withdrawn.

Germany itself was divided into four zones of occupation: British, American, French and Soviet. Berlin was also partitioned between the four states, though it was within the Soviet zone. In July 1945 an Allied Control Commission was established to oversee the occupation, whilst each zone established its own military and civilian authorities. There had been much talk of dividing Germany into separate states during the war, but despite renewed French interest in controlling the Saar, which remained separate from the rest of the German area until 1955, and in establishing an independent Rhineland state, the only permament division was the one that developed between the Western and Soviet zones as mutual political distrust and hostility grew.

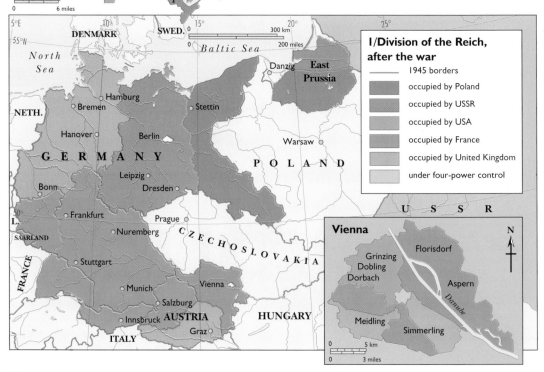

2/Administrative regions of occupied Germany, 1949

- occupied by Western powers
- occupied by Soviet Union

Unable to agree to a single German state the zonal authorities began to re-establish political and economic life in their respective zones. In 1947 Britain and the United States merged their zones to form Bi-Zonia. Within the western zones German political parties were allowed only if they were committed to parliamentary democracy. A Christian-Democratic Union had been formed in December 1945, based on the old Catholic Centre Party. The SPD became the main left-wing party. In 1949, in the absence of any agreement with the USSR over the future of Germany, the western states allowed free elections in their zones and in September 1949 Konrad Adenauer became Chancellor of a new Federal Republic of Germany. A few weeks later the Soviet zone became a separate German Democratic Republic, dominated by the communist-led Socialist Unity Party, and Germany was effectively partitioned until the collapse of Soviet communism forty years later.

One of the most pressing problems faced by the new Federal Republic was the vast influx of German refugees from the east who were either expelled or sought asylum. Some 6 million arrived between 1945 and 1949 with few possessions and no home. The process of rehabilitation went on to the 1960s.

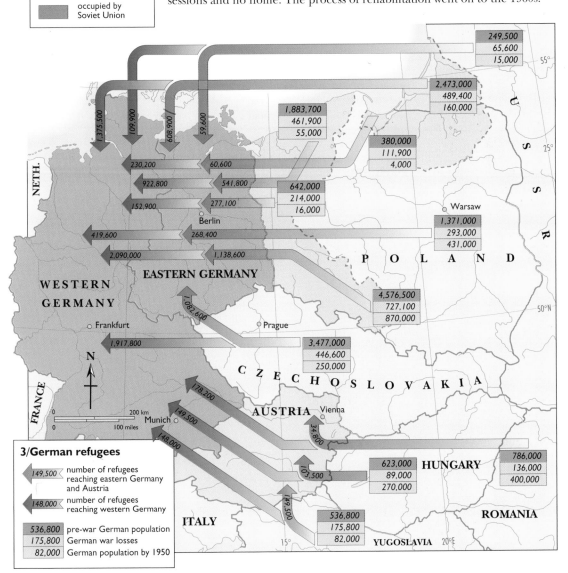

3/German refugees

- 149,500 number of refugees reaching eastern Germany and Austria
- 148,000 number of refugees reaching western Germany

536,800	pre-war German population
175,800	German war losses
82,000	German population by 1950

The Costs and Consequences of Defeat

For Germany the initial consequences of defeat were catastrophic. Germans called 1945 Year Zero, the moment of defeat 'Hour Zero'.

"I don't care what happens to the German population...I would take every mill and factory and wreck it."
Henry Morgenthau, September, 1944

The immediate consequences of defeat for Germany were catastrophic. Millions of German refugees and displaced persons cluttered German roads. More than three million dwellings were completely destroyed by the bombing. Industrial production came virtually to a halt, and revived only slowly in those sectors permitted by the occupation authorities. More than 10 million men were prisoners-of-war. They returned slowly from captivity but very large numbers in both Soviet and western hands died of disease or malnutrition. The deaths of more than three and a half million German soldiers during the war reduced the male population of some year-groups by as much as 38%. In 1950 the female population of the Federal Republic exceeded the male by three million. During the early years after the war the majority of the German workforce was made up of women and juveniles.

The occupying powers carried out a programme of physical reparation to add to German economic difficulties. Most of the dismantled plants went to the Soviet Union, but a good deal of technical and scientific equipment was taken in the western zones, along with a number of key research scientists. In 1946 a Level of Industry plan was introduced to limit German industrial development and to

1/Daily diet in Germany 1939–47

	Calories
1939/40	2,435
1940/41	2,445
1941/42	1,928
1942/43	2,078
1943/44	1,981
1944/45	1,671
1945/46	1,412
1946/47	1,421

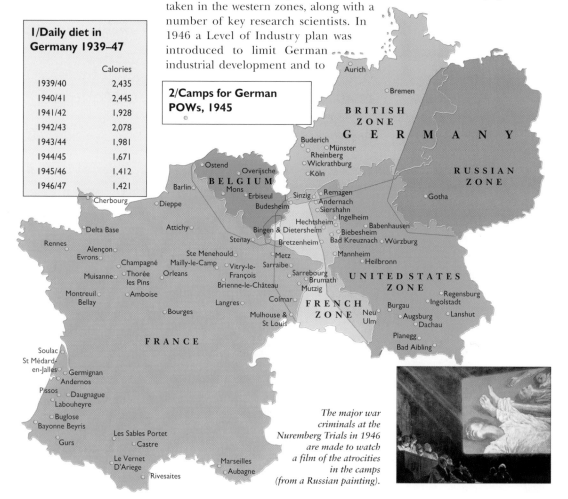

2/Camps for German POWs, 1945

The major war criminals at the Nuremberg Trials in 1946 are made to watch a film of the atrocities in the camps (from a Russian painting).

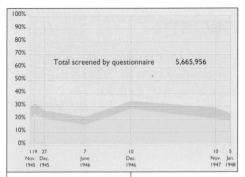

**3/Denazification:
British Zone 1945–49**

Question: *"Do you think that
the entire German people are
responsible for the war
because they let a government
come to power which plunged
the whole world into war?"*

yes

no

no opinion

prevent the revival of war-related sectors such as chemicals or aviation. Food supply was poor as German farmers lacked essential resources to expand production. Supplies from the west were intermittent, leading during 1946 and 1947 to severe shortages, bordering at times on starvation. The German diet, particularly in the cities, remained the monotonous wartime diet of bread, potatoes and small quantities of meat. Not until the stabilization of the currency and the end of inflation in 1948 did production of food and consumer products revive. By 1950 the output level of 1936 was reached, though living standards for many Germans, particularly among the refugee communities, remained low throughout the 1950s.

The occupying powers were also determined to bring German leaders to trial, and to impose a programme of de-Nazification. The main defendants were arraigned at Nuremberg in November 1945 for crimes against peace, against humanity, and against the laws of war. In October 1946 10 out of 21 were executed. Thousands of other Nazi and SS or police officials were murdered at the war's end or were tried and executed in further trials. Hundreds of thousands of former Nazis were sacked from public office and millions were forced to fill in de-Nazification questionnaires. In the end the chaos this produced brought the programme to an end and many former Nazis were re-employed. Opinion polling in the 1940s showed that many Germans still saw Nazism as a good idea poorly executed.

**5/German POWs
in Allied hands**

Total captured

In Northwest Europe	7,294.539
In Italy/Austria	1,425,000
In North Africa	371,000
In Eastern Europe	2,388,000
Wartime total	11,428,839

**4/Percentage deaths
of male population**

total deaths

missing presumed
dead

killed in war

5.86%   percentage of total

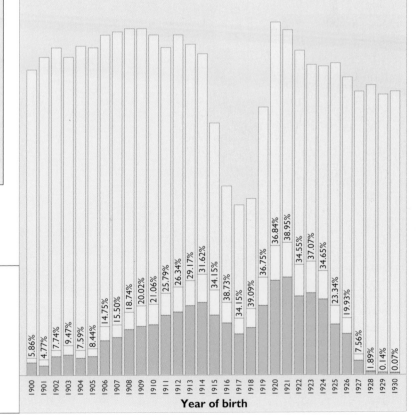

The Survival of Neo-Nazism

From the founding of West Germany in 1949 onwards small ultra-nationalist parties, many with direct links back to the NSDAP, have tried to revive German racism and chauvinism, with limited success.

"The Germans in the east, the centre and the west are one race, one nation... We have the right to live together as a united German nation in a single German state."
NPD leader, 1989

Despite the Allied programme of de-Nazification in the five years following the end of the war, radical nationalist parties revived quickly. In 1946 the German Rights Party (DRP) campaigned on a pseudo-Nazi programme, and won 5 seats in the new Federal Parliament in 1949. A second party, the Socialist Reich Party, formed in 1949, was led by former Nazis and adopted fascist rhetoric on issues of authority and national revival. It attracted some 10,000 members and won 11% of the vote in regional elections in Lower Saxony. In 1952 it was banned by the West German Constitutional Court because of its anti-democratic credentials.

Support for radical nationalism came not only from former Nazis, many of whom found themselves in positions of authority again in the 1950s, but from the millions of expellees from the east who were fiercely anti-commu-

Ultra-nationalist and Neo-nazi parties 1965-86	
Members	
1965	26,100
1966	34,500
1967	38,500
1968	36,800
1969	36,300
1970	29,500
1971	27,700
1972	24,500
1973	21,500
1974	21,200
1975	17,300
1976	20,300
1977	18,200
1978	17,800
1979	17,600
1980	19,800
1981	20,300
1982	19,000
1983	20,300
1984	22,100
1985	22,100
1986	22,100

3/Neo-nazism and racism, 1989–90
recorded Neo-nazi activities, 1989
racist excesses, 1989

The badge of modern nationalism reads 'I am proud to be a German'. Modern movements of the extreme right have revived the emblems of an earlier age.

nist and wanted a restoration of Germany's pre-war frontiers. This ambition became the stock-in-trade of all extreme nationalist parties founded in the post-war era. The most successful was the National Democratic Party (NPD), founded in 1964 by remnants of the failed DRP and the main expellee party, the BHE. Led by Adolf von Thadden, the NPD won 12 seats in the Baden-Württemberg parliament in 1968, and 14 in Bavaria in 1966. In the 1969 Federal election, following three years of intensive nationalist agitation, the Party polled 1.4 million votes. This was 4.3% of the whole, just short of the 5% necessary to gain seats in the Bundestag under the system of proportional representation.

During the 1970s radical nationalism declined as a political force as West Germany prospered and a new generation of young Germans who could not remember the Third Reich began to enter national politics. The economic slow-down, and the influx of political refugees from overseas and from eastern Europe, produced a neo-Nazi revival in the 1980s. The most overtly Nazi, the National Socialist Action Front, founded by Michael Kühnen in 1977, was banned in 1983, but there were plenty of imitators. By the late 1980s neo-fascist violence, particularly attacks on racial minorities, was widespread. In 1986 there were 1,281 prosecutions for right-wing violence. In 1992 there were 2,500 reported racist attacks and 17 deaths. There is little about the current economic or political context in Germany to suggest a revival of mass fascism, but the ghost of Hitlerism has yet to be laid.

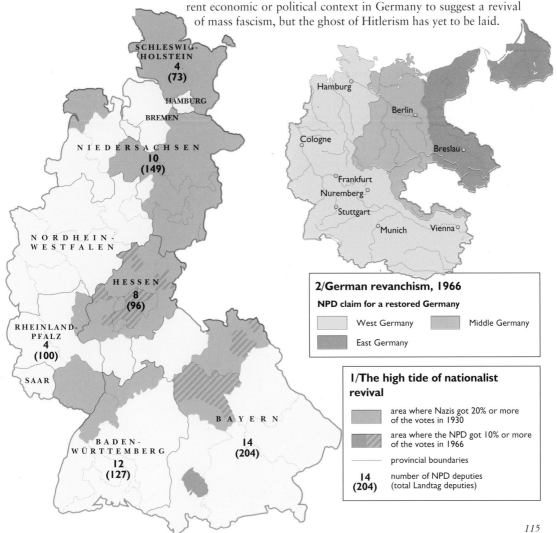

SCHLESWIG-HOLSTEIN
4
(73)

HAMBURG

BREMEN

NIEDERSACHSEN
10
(149)

NORDHEIN-WESTFALEN

HESSEN
8
(96)

RHEINLAND-PFALZ
4
(100)

SAAR

BAYERN
14
(204)

BADEN-WÜRTTEMBERG
12
(127)

Hamburg

Berlin

Cologne

Breslau

Frankfurt
Nuremberg

Stuttgart

Munich   Vienna

2/German revanchism, 1966

NPD claim for a restored Germany

	West Germany		Middle Germany
	East Germany		

1/The high tide of nationalist revival

area where Nazis got 20% or more of the votes in 1930

area where the NPD got 10% or more of the votes in 1966

provincial boundaries

**14
(204)** number of NPD deputies (total Landtag deputies)

Timeline: 1914–1933

GERMAN POLITICS	WAR AND FOREIGN POLICY	NAZI MOVEMENT	SOCIETY AND ECONOMY
Aug. 1914 'Burgfrieden' (political truce) declared	Aug. 1914 Outbreak of war with Russia, France and Britain 26 Aug. 1914 Battle of Tannenberg June 1916 Field Marshal Paul von Hindenburg becomes chief-of-staff	Aug. 1914 Hitler volunteers for military service Oct. 1916 Hitler wounded	Aug. 1914 Rathenau sets up Raw Materials Office June 1916 Hindenburg Munitions Programme launched Dec. 1916 Auxiliary Service Law for labour conscription
July 1917 Reichstag Peace Resolution	1 Feb. 1917 unrestricted submarine warfare begins June 1917 German bombers attack London		
	3 Mar. 1918 Treaty of Brest-Litovsk March 1918 German 'Spring Offensive' on western front		Mar. 1918 strike wave in Berlin
3 Oct. 1918 Prince Max of Baden becomes chancellor 3 Nov. 1918 Kiel naval mutiny 9 Nov. 1918 Friedrich Ebert appointed chancellor, Republic declared Jan. 1919 Communist uprising 19 Jan. 1919 National Assembly elected (SPD 38%) April 1919 Communist revolution in Bavaria 31 July 1919 Weimar Constitution adopted for republic	11 Nov. 1918 armistice comes into force at German request 28 June 1919 Versailles Treaty signed	Oct. 1918 Hitler wounded and gassed Jan. 1919 Anton Drexler founds German Workers' Party (DAP) Sept. 1919 Hitler joins DAP	Dec. 1918 Spartacus League launches social revolution Feb. 1920 Works Councils become law
Mar. 1920 Kapp Putsch in Berlin April 1920 Civil war in Ruhr area		Apr. 1920 DAP changes name to National Socialist German Workers' Party (NSDAP)	
	March 1921 Silesian plebiscite	July 1921 Hitler becomes chairman of NSDAP Nov. 1921 SA (Sturmabteilungen) formed	Apr. 1921 Reparations bill fixed at 132 bill. gold marks
24 June 1922 Rathenau assassinated	16 April 1922 German-Soviet Treaty of Rapallo 11 Jan. 1923 French and Belgian troops occupy the Ruhr		
Oct. 1923 Communists overthrown in Saxony and Thuringia 9 Nov. 1923 Hitler Putsch in Munich 16 April 1925 Hindenburg elected president		Nov. 1923 Hitler arrested after coup failure in Munich 1 April 1924 Hitler jailed for 5 years 20 Dec. 1924 Hitler released early Feb. 1925 Party re-founded at congress in Bamberg	Nov. 1923 German currency collapses Dec. 1923 new Rentenmark currency comes into use 9 Apr. 1924 Dawes Plan revises reparations schedule

Timeline: 1914–1933

28 May 1928 Reichstag election (12 Nazi deputies)
May 1928 SPD-led coalition under Hermann Müller

May 1930 Heinrich Brüning (Centre Party) appointed Chancellor
14 Sept. 1930 Reichstag election (107 Nazi deputies)

10 April 1932 Hindenburg re-elected as President
May 1932 Franz von Papen (Centre Party) replaces Brüning
June 1932 Anti-Fascist Action Front set up
1 June 1932 Franz von Papen (Centre Party) replaces Brüning
31 July 1932 Reichstag elections (230 Nazi deputies)

6 Nov. 1932 Reichstag elections (196 Nazi deputies)
3 Dec. 1932 General Kurt von Schleicher becomes chancellor

30 Jan. 1933 Adolf Hitler becomes chancellor

1 Dec. 1925 Locarno Treaties signed

8 Sept. 1926 Germany enters League of Nations

27 Aug. 1928 Germany signs Kellogg Pact
29 Sept. 1928 'First Armament Programme' approved

3 Oct. 1929 Gustav Stresemann dies
Jan. 1930 last Allied troops leave Rhineland

Jan. 1932 'Second Armaments Programme' authorized
2 Feb. 1932 Disarmament Conference opens in Geneva

June 1932 Lausanne Conference suspends reparations

14 Sept. 1932 Germany leaves Disarmament Conference (returns 11 Dec.)

July 1925 *Mein Kampf* published

Dec. 1926 Joseph Goebbels appointed Gauleiter of Berlin

6 Jan. 1929 Heinrich Himmler appointed head of SS (Schutzstaffel)

Sept. 1930 Office for Agriculture set up under Walther Darré

25 Feb. 1932 Hitler given German citizenship
13 Mar. 1932 Hitler fails to win presidential election

Dec. 1932 Gregor and Otto Strasser split with Hitler and leave party

Nov. 1928 steel industry lock-out

May 1929 Young Plan for reparations negotiated

6 Aug. 1929 Young Plan comes into force

Oct. 1929 Wall Street crash

20 June 1931 Hoover Moratorium on German foreign debts
13 July 1931 German banking crisis
8 Dec. 1931 Brüning introduces austerity measures and cuts wages

4 Sept. 1932 Papen work-creation programme launched

Dec. 1932 'Sofort-Programm' for work-creation

Jan. 1933 Unemployment reaches its highest point

Timeline: 1933–1939

GERMAN POLITICS	WAR AND FOREIGN POLICY	NAZI MOVEMENT	SOCIETY AND ECONOMY
27 Feb. 1933 Reichstag Fire	Feb. 1933 Disarmament Conference meets in Geneva (Germany leaves it in October 1933)	Feb. 1933 SS and SA recruited as auxiliary policemen	
28 Feb. 1933 Hitler granted emergency powers		Feb. 1933 first concentration camp set up at Oranienburg	16 March 1933 Hjalmar Schacht appointed President of the Reichsbank
5 March 1933 Reichstag elections (288 Nazi deputies)			
24 March 1933 Enabling Law passed			
31 March 1933 *Länder* lose their autonomy			
26 Apr. 1933 Gestapo founded		1 May 1933 new membership of Party suspended	2 May 1933 Trade Union organization and assets taken over by the state
26 May 1933 Law for the Seizure of Communist Assets			10 May 1933 German Labour Force (DAF) set up
22 June 1933 SPD banned			15 May 1933 Reich Hereditary Farm Law
14 July 1933 Law against the Establishment of New Parties			19 May 1933 Reich Labour Trustees appointed
		9 June 1933 Sicherheitsdienst (SD) established	1 June 1933 Law for the Reduction of Unemployment (Reinhardt-Programme for work-creation)
		17 June 1933 Baldur von Schirach appointed Reich Youth Leader	
		26 June 1933 Law for the Prevention of Hereditarily Diseased Progeny	
	July 1933 secret German Air Force established		
	20 July 1933 Concordat signed with Vatican		
			Sept. 1933 Hitler approves *Reichsautobahnenen* scheme
			27 Sept. 1933 Ludwig Müller elected Reich Bishop by the Evangelical Church
12 Nov. 1933 plebiscite and Reichstag elections	14 Oct. 1933 Germany withdraws from the League of Nations	Nov. 1933 Reichstag elections produce 93% Nazi vote	
30 Jan. 1934 Law on the Reconstruction of the Reich	26 Jan. 1934 Non-Aggression Pact signed with Poland		16 Jan. 1934 Law for the Organization of National Labour
		30 June 1934 Röhm and SA leaders purged in Night of the Long Knives	20 July 1934 New Plan established for German trade
	25 July 1934 Engelbert Dollfuss murdered by Austrian Nazis		
2 Aug. 1934 death of President Hindenburg	2 Aug 1934 armed forces swear oath of loyalty to Hitler	2 Aug. 1934 Hitler declares himself Führer, merging offices of Chancellor, President and Party Leader	
19 Aug. 1934 plebiscite on Hitler becoming President and chancellor			Sept. 1934 Schacht becomes Minister of Economics
			Nov. 1934 Carl Goerdeler appointed Price Commissioner
30 Jan. 1935 remaining *Länder* powers taken over by central government	13 Jan. 1935 Saar plebiscite (91% favour union with Germany)		
	16 March 1935 conscription re-introduced and rearmament publicly announced		
Apr. 1935 Reichs justice system centralized	11 Apr. 1935 Stresa Front meeting		
	18 June 1935 Anglo-German Naval Agreement		26 June 1935 Reich Labour Service Law
14 Nov. 1935 Reich Citizenship Law		15 Sept. 1935 Race legislation announced at the Nuremberg Party rally	
	7 March 1936 Germany re-militarizes the Rhineland	10 Feb. 1936 concentration camps come formally under SS control	

Timeline: 1933–1939

26 Jan. 1937 Law on
German Civil Servants

Feb. 1938 Blomberg-Fritsch
crisis
5 Feb. 1938 last meeting of
full Reich cabinet

10 Apr. 1938 plebiscite on
Anschluss with Austria (99%
in favour)

14 April 1939 'Ostmark
Law' integrates Austria with
German state
administration

Aug 1936 Olympic games
staged in Berlin

25 Nov. 1936 German-
Japanese Anti-Comintern
Pact

5 Nov. 1937 'Hossbach'
conference on Hitler's
future strategy
6 Nov. 1937 Italy joins Anti-
Comintern Pact
3 Feb. 1938 Hitler assumes
Supreme Command of the
armed forces and
establishes OKW
12 March 1938 German
troops occupy Austria and
Anschluss declared

29 Sept. 1938 Munich
agreement over the Czech
crisis

Jan. 1939 German Navy 'Z-
Plan' launched
15 March 1939 German
Protectorate established
over Bohemia and Moravia
23 March 1939 Memel re-
occupied
22 May 1939 Pact of Steel
signed with Italy

23 Aug. 1939 German-Soviet
Non-Aggression Pact signed
in Moscow
1 Sept 1939 Germany
invades Poland
3 Sept 1939 Britain and
France declare war on
Germany

---

24 April 1936 Party training
institutes set up under DAF
17 June 1936 Himmler
becomes head of all police
and security services

1 Dec. 1936 Hitler Youth
becomes an agency of the
state

4 Feb. 1938 Joachim von
Ribbentrop becomes
Foreign Minister

1 Sept. 1938 last Party
Congress at Nuremberg
9 Nov. 1938 'Kristallnacht'
pogrom against German
Jews

1 May 1939 Nazi
membership roles re-
opened
1 May 1939 Konrad Henlein
appointed Gauleiter of
Sudetenland
May 1939 Party Gau
organization set up in
Austria

27 Sept. 1939
Reichssicherheitshauptamt
(RSHA) set up

---

9 Sept. 1936 Four-Year Plan
announced at Nuremberg
Rally
19 Oct. 1936 Göring takes
over as head of Four Year
Plan
Oct. 1936 Reich Office for
combating Homosexuality
and Abortion set up
26 Nov. 1936 Price-Stop
Decree

July 1937 Reichswerke
'Hermann Göring' set up
Nov. 1937 Schacht resigns as
Minister of Economics

4 Feb. 1938 Walther Funk
appointed as Economics
Minister

26 May 1938 Hitler launches
Volkswagen project

3 Dec. 1938 Decree on the
Compulsory Aryanization of
Jewish businesses
20 Jan. 1939 Funk becomes
President of Reichsbank
March 1939 Germany takes
control of Czech Skoda
armaments works

April 1939 Funk announces
New Finance Plan to fund
rearmament

4 Sept. 1939 War Economy
Decree comes into force

Timeline: 1939–1945

GERMAN POLITICS	WAR AND FOREIGN POLICY	NAZI MOVEMENT	SOCIETY AND ECONOMY
		1 Sept. 1939 14 Gauleiter appointed as Reich Defence Commisioners	4 Sept. 1939 comprehensive rationing introduced in Germany
6 Oct. 1939 Hitler offers peace to Britain and France in Reichstag speech	27 Sept 1939 fall of Warsaw 28 Sept. 1939 Partition of Poland agreed with USSR	8 Nov. 1939 unsuccessful attempt on Hitler's life	
8 Oct. 1939 western Polish territories incorporated into German Reich		Nov. 1939 Waffen–SS established	Oct. 1939 Hitler formally endorses 'euthenasia' programme
			Dec. 1939 Hitler authorizes 'Armament 1940–42 programme' to double military output
			Jan. 1940 increases in corporation, luxury and income taxes
			17 March 1940 Fritz Todt appointed Minister of Munitions
	9 Apr. 1940 Germany invades Denmark and Norway		
	10 May 1940 Germany invades Netherlands, Belgium, Luxembourg and France	May 1940 Julius Streicher, Gauleiter of Franconia, dismissed for corruption	
	14 May 1940 bombing of Rotterdam		
	22 June 1940 France signs armistice		
19 July 1940 Hitler offers Britain peace on German terms	July 1940 start of Battle of Britain	July 1940 Göring promoted to rank of Reich Marshal	July 1940 Funk launches the economic New Order
27 Sept. 1940 Germany signs Tri-Partite Pact with Italy and Japan	7 Sept. 1940 'Blitz' against British cities begins nine months of bombing		
12 Oct. 1940 Alsace Lorraine, Luxembourg brought under direct German rule by Reich commissioners	12 Feb. 1941 Rommel sent to North Africa with *Afrika Korps*		
	6 Apr. 1941 Germany invades Yugoslavia and Greece		
	27 May 1941 sinking of the *Bismarck*	10 May 1941 Hitler's deputy, Rudolf Hesse, flies to Scotland	May 1941 Hitler issues his first decree on rationalizing the war economy
6 June 1941 'Commissar Order' on liquidating political officials in the USSR	22 June 1941 start of operation Barbarossa against the USSR	12 May 1941 Martin Bormann becomes Hitler's deputy	
July 1941 Hitler orders the liquidation of the Jews	July 1941 900-day siege of Leningrad begins	17 July 1941 Himmler granted special security powers for the occupied east	6 July 1941 Göring aircraft programme launched to expand aircraft production
		17 July 1941 Alfred Rosenberg appointed Minister for the Eastern Territories	27 July 1941 Göring establishes the economic organization for the occupied Soviet territories
		31 July 1941 Göring orders Reinhard Heydrich to find a 'Final Solution' to the Jewish question	
1 Sept. 1941 all Jews forced to wear a yellow star	23 Sept. 1941 Germans capture Kiev	20 Aug. 1941 Erich Koch, Gaulieter of East Prussia, appointed Reich Commissar of the Ukraine	Sept. 1941 the Soviet iron and steel industry reorganized under the German monopoly BHO organization
3 Oct. 1941 Hitler announces victory over the USSR and start of the European New Order			
5 Oct. 1941 start of the transfer of German Jews to the east			
	11 Dec. 1941 Germany declares war on the USA	19 Dec. 1941 Hitler assumes personal command of the German army	3 Dec. 1941 Hitler's 'Rationalization Decree' published for reorganizing the war economy

Timeline: 1939–1945

20 Jan. 1942 Wannsee Conference at which plans for Jewish genocide are confirmed

April 1942 Hitler wins special judicial authority from the Reichstag

20 Aug. 1942 Otto Thierack appointed Minister of Justice
20 Aug. 1942 Roland Freisler appointed to head People's Court

18 Feb. 1943 student uprising in Munich by 'White Rose' organization
18 Feb. 1943 Goebbels delivers 'Total War' speech in Berlin
April 1943 Goebbels attempt to reform political power structure

Jan. 1944 Kreisau resistance circle broken up

25 July 1944 Goebbels appointed Reich Plenipotentiary for Total War

9 Apr. 1945 senior resistance figures executed at Flossenburg
1 May 1945 Admiral Dönitz becomes head-of-state

5 June 1945 Allied Control Commission formally established
17 July 1945 Potsdam conference opens in Berlin
1 Jan. 1947 Bi-Zonia created out of US and British zones
23 May 1949 German Federal Republic founded
15 Sept. 1949 Konrad Adenauer becomes first Chancellor of Federal Republic
7 Oct. 1949 Soviet-backed German democratic Republic established

30–31 May 1942 1,000-bomber raid on Cologne
28 June 1942 launch of Operation Blue in southern USSR

4 Nov. 1942 Rommel defeated at el Alamein
11 Nov. 1942 Germany occupies Vichy France
19 Nov. 1942 Red Army launches counter-offensive around Stalingrad
31 Jan. 1943 Paulus surrenders in Stalingrad

7 May 1943 destruction of the Warsaw ghetto
13 May 1943 German forces surrender in Tunisia
24 May 1943 Dönitz orders submarines to abandon the Battle of the Atlantic
5 July 1943 start of the Battle of Kursk (Operation Citadel)
6 Nov. 1943 Germans lose Kiev

March 1944 German forces occupy Hungary
6 June 1944 western Allies invade Normandy
12 June 1944 first VI lands on London
1 Aug. 1944 Warsaw rising by Polish resistance
8 Sept. 1944 first V2 rocket falls on London
16 Dec. 1944 start of the second Ardennes offensive (Battle of the Bulge)
14 Jan. 1945 Soviet forces reach east Prussia
7 March 1945 western Allies cross Rhine
13 Apr. 1945 Vienna occupied by Red Army
23 Apr. Soviet forces reach Berlin
2 May 1945 German forces surrender in Italy
7 May 1945 General Jodl signs instrument of surrender

March 1942 SS begin Operation Reinhard for the liquidation of the Polish Jews

4 June 1942 Heydrich assassinted in Prague
23 June 1942 first mass gassings at Auschwitz

16 Nov. 1942 all 42 Gauleiter appointed Reich Defence Commissioners

24 Aug. 1943 Himmler becomes Minister of Interior

20 July 1944 Stauffenberg attempts to assassinate Hitler at his HQ in East Prussia

Apr. 1945 Göring attempts to usurp Hitler's authority and is expelled from office
30 Apr. 1945 Hitler commits suicide in Berlin. Goebbels kills himself and his family

20 Nov. 1945 opening of the Nuremberg Trials
16 Oct. 1946 execution of major war criminals, including Ribbentrop, Rosenberg, Frick, Frank, Streicher and Sauckel

20 Dec. 1941 Fritz Todt sets up war production committees
9 Feb. 1942 Albert Speer appointed Minister of Weapons and Munitions
21 March 1942 Fritz Sauckel, Gauleiter of Thuringia, appointed Plenipotentiary for Labour
April 1942 Speer establishes Central Planning agency

27 Jan. 1943 Civil conscription of women
4 Feb. 1943 inessential businesses ordered to close

26 June 1943 Speer takes over all war production except aircraft
24–25 July 1943 Hamburg heavily bombed and estimated 44,000 killed

Aug. 1943 Himmler begins the underground factory programme
March 1944 the 'Fighter Staff' for the emergency production of aircraft set up by Speer

25 September 1944 Hitler forms Volkssturm home defence units

13 Feb. 1945 Allies bomb Dresden, 50,000+ killed
19 March 1945 Hitler orders 'scorched earth' in Germany

Statistical Appendix

I: From War to Third Reich 1918–1933

Losses of Territory and Population under the Versailles Settlement

Territory	Area in km^2	Population	% German	Date area lost
Eupen-Malmédy	1,036	60,000	82.6	20.09.20
Alsace-Lorraine	14,522	1,874,000	87.4	10.01.20
North Schleswig	3,993	166,300	25.0	15.06.20
East Prussia (part)	501	24,700	40.9	10.01.20
West Prussia (part)	5,864	964,700	44.2	10.01.20
Posen	26,042	1,946,400	35.0	10.01.20
Pomerania & Brandenburg (part)	10	200	90.0	10.01.20
Lower Silesia (part)	511	26,200	44.7	10.01.20
South and east Upper Silesia	3,221	892,000	29.4	19.06.22
Memelland	2,656	141,200	52.3	10.01.20
Danzig	1,914	330,600	96.3	10.01.20
Hultschinderland (Upper Silesia)*	315	48,500	14.8	10.01.20
Saarland (under League contol)	1,922	651,900	99.7	10.01.20

* area of Upper Silesia lost to Czechoslovakia. The other territories lost in Silesia went to Poland.

Reparations Payments 1919–1932

1919–1924	German estimate	51.6 billion gold marks
	Allied valuation	8.0 billion gold marks

1925–1932 (billion RM)

	Allied requirement		Actual payment	as % of National Income
1925	1.00	Dawes	1.06	1.8
1926	1.22	Dawes	1.19	2.0
1927	1.50	Dawes	1.58	2.3
1928	2.50	Dawes	2.00	2.8
1929	1.94	Dawes	2.34	3.2
1930	1.70	Young	1.71	2.6
1931	1.69	Young	0.99	1.8
1932	1.73	Young	0.16	0.4

German Industrial Production 1913–1933 (1928 = 100)

1913	98	1925	81	1932	58
1919	37	1926	78	1933	66
1920	54	1927	98		
1921	65	1928	100		
1922	70	1929	100		
1923	46	1930	87		
1924	69	1931	70		

Inflation in Germany: Selected Statistics

Money Supply

Total cash in circulation, end of year (Dec): billion marks

1913	6.6	1919	50.1
1914	8.7	1920	81.6
1915	10.0	1921	122.9
1916	12.3	1922	1,295.2
1917	18.4	1923	496 trillion
1918	33.1		

The Effects of Inflation

Average real weekly wage (1913 = 100)

	Skilled railway workers	Ruhr miners	Senior civil servants
1914	97.2	93.3	97.2
1919	92.2	82.4	40.2
1921	74.5	89.1	39.3
1922	64.2	69.9	35.6
1923	50.9	70.1	38.0

The Party Share of the Vote in Reichstag Elections 1919–1933 (%)

Party	1919	1920	1924 (May)	1924 (Dec)	1928	1930	1932 (Jul)	1932 (Nov)	1933
KPD	–	2.1	12.6	9.0	10.6	13.1	14.5	16.9	12.3
USPD	–	7.6	17.9	0.8	0.3	0.1	0.0	–	–
SPD	–	37.9	21.7	20.5	26.0	29.8	24.5	21.6	20.4
DDP	–	18.6	8.3	5.7	6.3	4.9	3.8	1.0	1.0
Zentrum	–	15.9	13.6	13.4	13.6	12.1	11.8	12.5	11.9
BVP	–	3.8	4.2	3.2	3.8	3.1	3.0	3.7	3.4
DVP	–	4.4	13.9	9.2	10.1	8.7	4.7	1.2	1.9
DNVP	–	10.3	15.1	19.5	20.5	14.2	7.0	6.2	8.9
NSDAP	–	–	6.5	3.0	2.6	18.3	37.4	33.1	43.9
Others *	–	1.6	3.3	8.6	7.5	13.9	13.8	2.0	2.6

* mainly small right-wing parties representing peasant or small business interests.

KPD = German Communist Party; USPD = Independent German Social Democratic Party; SPD = German Social Democratic Party; DDP = German Democrats; Zentrum = Catholic Centre Party; BVP = Bavarian People's Party; DVP = German People's Party; DNVP = German Nationalist Party; NSDAP = National Socialist German Worker's Party (Nazis)

The Regional Distribution of Nazi Party votes 1924–33

(% of registered voters in the region cast for the NSDAP)

Election	North West	North East	West Central	East Central	South
1924 May	6	8	3	5	7
1924 Dec	3	5	1	2	3
1928	3	1	2	2	3
1930	18	18	14	15	13
1932 July	37	38	27	34	28
1932 Nov	30	31	23	29	24
1933	41	47	34	40	38

Composition of the Nazi Constituency in 1928–1933 (%)					
Election	1928	1930	1932 (J)	1932 (N)	1933
Catholics	30	20	17	17	24
Other Denominations	70	80	83	83	76
Comunity size					
0–5000	39	41	45	47	47
5–20,000	14	13	13	12	12
20–100,000	16	15	13	13	13
over 100,000	31	31	28	27	28
Social group					
Workers *	40	40	39	39	40
White-collar **	22	21	19	19	18
Farmers & independents ***	37	39	42	41	42

*   includes unskilled, skilled, and craft workers

**   includes civil servants, teachers, salaried officials

*** includes professions, small to large business, and farmers

Occupational Structure of the German Population, Summer 1933	
Unskilled workers	10,075,782
Skilled workers	4,478,803
White-collar employees	3,359,248
Master craftsmen	2,585,551
Farmers	2,082,912
Merchants	1,624,118
Civil Servants	1,530,983
Professionals	742,518
Managers and Entrepreneurs	234,955

II: Establishing the Dictatorship

Membership of Nazi Youth Organizations 1932–1939 (in millions)					
Date	Hitler Youth	Jungvolk	Bund deutscher Mädel	Jung–Mädel	Total
1932	0.05	0.02	0.02	0.004	0.1
1933	0.56	1.13	0.24	0.35	2.29
1934	0.78	1.46	0.47	0.86	3.58
1935	0.83	1.50	0.57	1.05	3.94
1936	1.17	1.78	0.87	1.61	5.44
1937	1.24	1.88	1.04	1.72	5.88
1938	1.66	2.06	1.45	1.86	7.03
1939	1.72	2.14	1.50	1.92	7.73
figures for 31 December 1932–38, 1939 figures for spring					

Total Nazi Party Membership	
Date	Number
1919	55
1921	3,000
1923	55,287
1928	96,918
1930	129,563
1933	849,000
1935	2,493,890
1937	2,793,890
1938	4,985,400
1939	5,339,567
1942	7,100,000
1943	7,600,000
1945	c.8,000,000

Work-creation Funds and Employment, 1932–1935			
Date	Expenditure (million RM)	Date	New jobs
1932–3	1,455	Jan. 1933	23,665
		July 1933	140,126
		Nov. 1933	400,847
		Jan. 1934	385,275
		Mar. 1934	630,163
1934	1,985	June 1934	392,433
1935	593		

Estimates of German Gross National Product 1928–1943 (billion RM)

Date	GNP at current prices	GNP at constant prices (1928–38 at 1928 prices; 1939–43 at 1939 prices)
1928	89.5	90.8
1929	89.7	88.5
1930	83.9	83.8
1931	70.4	76.1
1932	57.6	71.9
1933	59.1	73.7
1934	66.5	83.7
1935	74.4	92.3
1936	82.6	101.2
1937	93.2	114.2
1938	104.5	126.2
1939	129.0	129.0
1940	132.0	129.0
1941	137.0	131.0
1942	143.0	136.0
1943	160.0	150.0

figures from 1939 include Austria, Sudetenland and Memel

Wages, Earnings, and Cost of Living, 1929–1938

	Money wages (1913/14 = 100)	Real wage rates	Real earnings (1925/9 = 100)	Cost-of-living index (1913/14 = 100)	Wholesale prices (1925 = 100)	Wages as % of National Income	Private consumption as % of National Income
1929	177	115	107	154.0	96.8	62	71
1932	144	120	91	120.6	68.1	64	83
1933	140	119	87	118.0	65.8	63	81
1934	140	116	88	121.1	69.3	62	76
1935	140	114	91	123.0	71.8	61	71
1936	140	112	93	124.5	73.4	59	64
1937	140	112	96	125.1	74.7	58	62
1938	141	112	101	125.6	74.6	57	59

Average Weekly Hours Worked in German Industry 1929–44

Date	All Industries	Producer Goods	Consumer Goods
1929	46.0	46.3	45.7
1933	42.9	43.0	42.9
1935	44.4	45.9	42.6
1937	46.1	47.3	44.5
1938	46.5	47.8	44.9
1939	47.6	48.2	45.9
1940	47.6	48.5	43.8
1941	49.1	49.9	45.8
1942	48.7	49.6	45.0
1943	49.1	49.9	45.3
1944	48.3	49.2	43.3

Registered Unemployed, 1929–1940 (thousands)*

Month	1929	1930	1931	1932	1933	1934
Jan.	2,850.2	3,217.6	4,886.9	6,041.9	6,013.6	3,772.7
Feb.	3,069.7	3,365.8	4,971.8	6,128.4	6,000.9	3,372.6
Mar.	2,483.9	3,040.7	4,743.9	6,034.1	5,598.8	2,798.3
Apr.	1,711.6	2,786.9	4,358.1	5,739.0	5,331.2	2,608.6
May	1,349.8	2,634.7	4,052.9	5,582.6	5,038.6	2,528.9
June	1,260.0	2,640.6	3,953.9	5,475.7	4,856.9	2,480.8
July	1,251.4	2,765.2	3,989.6	5,392.2	4,463.8	2,426.0
Aug.	1,271.9	2,882.5	4,214.7	5,223.8	4,124.2	2,397.5
Sept.	1,323.6	3,004.2	4,354.9	5,102.7	3,849.2	2,281.8
Oct.	1,557.1	3,252.0	4,623.4	5,109.1	3,744.8	2,226.6
Nov.	2,035.6	3,698.9	5,059.7	5,355.4	3,714.6	2,352.6
Dec.	2,850.8	4,383.8	5,668.1	5,772.9	4,059.0	2,604.7
AVERAGE	1,898.6	3,075.5	4,519.7	5,575.4	4,804.4	2,718.3

Month	1935	1936	1937	1938	1939	1940
Jan.	2,973.5	2,520.4	1,853.7	1,051.7	301.8	159.7
Feb.	2,764.1	2,514.8	1,610.9	946.3	196.3	123.8
Mar.	2,401.8	1,937.1	1,245.3	507.6	134.0	66.2
Apr.	2,233.2	1,762.7	960.7	422.5	93.9	39.9
May	2,019.2	1,491.2	776.3	338.3	69.5	31.7
June	1,876.5	1,314.7	648.4	292.2	48.8	26.3
July	1,754.1	1,169.8	562.8	218.3	38.3	25.0
Aug.	1,706.2	1,098.4	509.2	178.7	33.9	23.1
Sept.	1,713.9	1,035.2	469.0	155.9	77.5	21.9
Oct.	1,828.7	1,177.4	501.8	163.9	79.4	–
Nov.	1,984.4	1,197.1	572.6	152.4	72.5	–
Dec.	2,507.9	1,478.8	994.7	455.6	104.4	–
AVERAGE	2,151.0	1,592.6	912.3	429.4	104.2	43.1

* from March 1935 including Saarland; from March 1939 including Sudetenland; from June 1939 including Memel.

Employment and Unemployment, by region, 1933–1937 ('000)

Region	Unemployment (end June)					Employment (end June)				
	1933	1934	1935	1936	1937	1933	1934	1935	1936	1937
East Prussia	76	14	7	4	4	453	535	543	562	576
Silesia	366	195	171	138	58	962	1,122	1,134	1,156	1,233
Brandenburg	752	406	249	175	96	1,805	2,104	2,265	2,371	2,557
Pomerania	86	24	23	12	7	418	509	484	510	529
Nordmark	338	192	127	85	53	867	1,008	1,102	1,174	1,242
Lower Saxony	272	112	67	29	8	891	1,086	1,134	1,238	1,316
Westphalia	390	196	157	114	46	1,093	1,298	1,365	1,462	1,584
Rhineland	667	393	373	276	151	1,499	1,754	1,837	2,152	2,335
Hessen	280	143	125	87	45	693	849	865	932	1,006
C. Germany	387	170	95	54	21	1,179	1,412	1,534	1,633	1,759
Saxony	595	321	265	199	97	1,314	1,551	1,592	1,676	1,811
Bavaria	394	199	140	94	37	1,395	1,640	1,724	1,832	1,950
SW Germany	254	116	78	48	26	1,104	1,300	1,349	1,425	1,520

Consumption in Working-Class Families 1927 and 1937 (annual)*

	1927	1937	% change
Rye bread (kg)	262.9	316.1	+ 20.2
Wheat bread (kg)	55.2	30.8	− 44.2
Meat and meat products (kg)	133.7	109.2	− 18.3
Bacon (kg)	9.5	8.5	− 10.5
Milk (ltr)	427.8	367.2	− 14.2
Cheese (kg)	13.0	14.5	+ 11.5
Eggs (number)	404.0	237.0	− 41.3
Fish (kg)	21.8	20.4	− 6.4
Vegetables (kg)	117.2	109.6	− 6.5
Potatoes (kg)	499.5	519.8	+ 4.1
Sugar (kg)	47.2	45.0	− 4.7
Tropical fruit (kg)	9.7	6.1	− 37.1
Beer (ltr)	76.5	31.6	− 58.7

* Adjusted for changes in purchasing-power and family size. Includes family budgets of low-paid civil servants and salaried workers.

Statistics on German Finance 1928/9–1938/9 (billion RM)*

Date	Government revenue	Government expenditure	Total debt	Money supply
1928–9	9.0	13.0	–	16.4
1932–3	6.6	9.2	12.3	13.4
1933–4	6.8	8.9	13.9	13.9
1934–5	8.2	12.6	15.9	15.7
1935–6	9.6	14.1	20.1	16.7
1936–7	11.4	17.3	25.8	18.1
1937–8	13.9	21.4	31.2	20.0
1938–9	17.7	32.9	41.7	23.7

* Fiscal year beginning 1 April for columns 1 and 2. End of fiscal year for column 3. End of calendar year for column 4.

III: Foreign Policy in Germany 1933–1939

German Foreign Trade 1928–38 (current prices, million RM)

Date	Imports	Exports	Balance
1928	14001.3	12275.6	−1725.7
1932	4666.5	5739.2	1072.7
1933	4203.6	4871.4	667.8
1934	4451.0	4166.9	−284.1
1935	4158.7	4269.7	111.0
1936	4217.9	4768.2	550.3
1937	5468.4	5911.0	442.6
1938*	6051.7	5619.1	−432.6

* inc. Austria

German Trade with the Balkan Region 1933–1939 (million RM): I = imports E = exports

Country		1933	1934	1935	1936	1937	1938	1939
Bulgaria	I	31.3	33.7	41.4	57.6	71.8	95.7	110.0
	E	17.7	19.3	39.9	47.6	68.2	61.6	97.8
Greece	I	53.4	55.3	58.5	68.4	76.4	101.0	92.1
	E	18.7	29.3	49.1	63.5	113.1	121.2	85.5
Yugoslavia	I	33.5	36.3	61.4	75.2	132.2	172.2	131.5
	E	33.8	31.5	36.9	77.2	134.4	144.6	181.3
Romania	I	46.1	59.0	79.9	92.3	179.5	177.8	209.5
	E	46.0	50.9	63.8	103.6	129.5	168.6	216.7
Hungary	I	34.2	63.9	77.9	93.4	114.1	186.2	222.5
	E	38.1	39.6	62.9	83.0	110.5	146.4	228.7

IV: Expansion and War 1939–1945

German Naval Construction 1939–1945

Ships	1939	1940	1941	1942	1943	1944	1945
Pocket battleships	3	–	–	–	–	–	–
Battle cruisers	2	–	–	–	–	–	–
Battleships	–	1	1	–	–	–	–
Cruisers	8	1	–	–	–	–	–
Destroyers	22	2	5	5	2	1	–
Torpedo boats	20	9	4	4	6	5	1
Submarines	58	68	129	282	207	258	139
Speed boats	20	26	34	44	46	52	11
Minesweepers	44	14	25	31	38	29	8

German Aircraft Production 1933–1945

Aircraft	1939	1940	1941	1942	1943	1944
Fighters	1,856	3,106	3,732	5,213	11,738	28,926
Bombers	2,877	3,997	4,350	6,539	8,589	6,468
Transport	1,037	763	969	1,265	2,033	1,002
Trainers	1,112	1,328	889	1,170	2,076	3,063
Others*	1,413	1,632	1,836	1,369	1,091	348
Total	8,295	10,826	11,776	15,556	25,527	39,807

* includes liaison aircraft, reconnaissance aircraft and other special duty or experimental aircraft

German Armed Forces, Stength and Losses 1939–1944 (millions)			
Date	Active Strength	Cumulative Losses	Total Mobilized
1939, May 31	1.4	–	1.4
1940, May 31	5.6	0.085	5.7
1941, May 31	7.2	0.185	7.4
1942, May 31	8.6	0.800	9.4
1943, May 31	9.5	1.7	11.2
1944, May 31	9.1	3.3	12.4
1944, Sept 30	9.1	3.9	13.0

Indices of German Weapons Output 1942–43 (Jan–Feb 42 = 100)			
Weapons	Annual average		
	1942	1943	1944
Total weapons	142	222	277
Aircraft	133	216	277
Ammunition	166	247	306
Guns	137	234	348
Armoured vehicles	130	330	536
Naval vessels	142	181	166
Explosive	129	199	212

Output of Major Categories of Armaments 1939–1945							
Armaments	1939	1940	1941	1942	1943	1944	1945
Aircraft	8,295	10,247	11,776	15,409	24,807	39,807	7,540
Aero-engines	3,865	15,510	22,400	37,000	50,700	54,600	–
Tanks		2,200	5,200	9,300	19,800*	27,300*	–
Munitions ('000 tons)		865	540	1,270	2,558	3,350	–
Automatic weapons ('000)		171	325	317	435	787	–
Heavy artillery and Flak ('000s)		6	30	69	157	361	–
* includeds self-propelled artillery							

V: The German New Order

Total Food Supplies from the Occupied Areas of the USSR 1941–1944 ('000 tonnes)				
Commodity	German Armed Forces	Export to Reich	German civilian admin. in USSR	Total
Total grains	4,050.0	1,760.8	3,341.1	9,151.9
Feed grain	1,828.3	972.3	1,334.9	4,135.5
Bread grain	2,221.7	788.5	2,006.2	5,016.4
Hay	1,816.5	–	691.0	2,507.6
Meat	411.9	66.6	85.1	563.7
Fish	43.7	1.1	22.9	67.6
Potatoes	2,039.6	13.0	1,229.1	3,281.7
Butter	118.1	20.8	67.9	206.8
Beet sugar	243.6	62.0	95.4	401.0

The Occupational Distribution of Foreign Labour in Germany in August 1944

	Agriculture	Mining	Metal	Chemicals	Building	Transport	Total
Belgians							
TOTAL	28,652	5,416	95,872	14,029	20,906	12,576	177,451
Civilians	3,948	2,787	86,441	13,533	19,349	11,585	137,643
POWs	24,704	2,629	9,431	496	1,557	991	39,808
as % of all							
Belgians workers	16.1	3.0	54.0	8.0	11.8	7.0	100%
French							
TOTAL	405,897	21,844	370,766	48,319	59,440	48,700	954,966
Civilians	54,590	7,780	292,800	39,417	36,237	34,905	465,729
POWs	351,307	14,064	77,966	8,902	23,203	13,795	489,237
as % of all							
French workers	42.5	2.3	39.0	5.0	6.2	5.0	100%
Italians							
TOTAL	45,288	50,325	221,304	35,276	80,814	35,319	468,326
Civilians	15,372	6,641	41,316	10,791	35,271	5,507	114,898
POWs	29,916	43,684	179,988	24,485	45,543	29,812	353,428
as % of all							
Italian workers	9.7	10.7	47.3	7.5	17.3	7.5	100%
Dutch							
Civilians	22,092	4,745	87,482	9,658	32,025	18,356	174,358
as % of all							
Dutch workers	12.7	2.7	50.2	5.5	18.4	10.5	100%
Soviet workers							
TOTAL	862,062	252,848	883,419	92,952	110,289	205,325	2,406,895
Civilians	723,646	92,950	752,714	84,974	77,991	158,024	1,890,299
POWs	138,416	159,898	130,705	7,978	32,298	47,301	516,596
as % of all							
Soviet workers	35.8	10.5	36.7	3.9	4.6	8.5	100%
Poles							
TOTAL	1,125,632	55,672	130,905	23,871	68,428	35,746	1,440,254
Civilians	1,105,719	55,005	128,556	22,911	67,601	35,484	1,415,276
POWs	19,913	667	2,349	960	827	262	24,978
as % of all							
Polish workers	78.1	3.9	9.1	1.7	4.7	2.5	100%
Bohemians, Moravians							
Civilians	10,289	13,413	80,349	10,192	44,870	18,566	177,679
as % of all							
citizens from the							
Protectorate	5.8	7.5	45.2	5.7	25.3	10.5	100%
TOTAL	2,499,912	404,263	1,870,097	234,297	416,772	374,588	5,799,929
Civilians	1,935,656	183,321	1,469,658	191,476	313,344	282,427	4,375,882
POWs	564,256	220,942	400,439	42,821	103,428	92,161	1,424,047
as %	43.1%	7.0%	32.2%	4.0%	7.2%	6.5%	100%

VI: German Society and Total War

Native Female Labour Force in Germany 1925–1944 (thousands)			
Date	Total	in Agriculture	in Industry
1925	11,478	4,970	2,988
1933	11,479	4,649	2,758
1939	12,701	4,880	3,310
1940	14,386	5,689	3,650
1941	14,167	5,369	3,677
1942	14,437	5,673	3,537
1943	14,806	5,665	3,740
1944	14,897	5,756	3,636

(figures for 1925–39 are annual averages, for 1940–43 are for May, figure for 1944 is for September)

Proportion of Women in the Total Native Civilian Workforce: Germany, Britain and the USA (%)			
Date	Germany	Britain	United States
1940	41.4	29.8	25.8
1941	42.6	32.2	26.6
1942	46.0	36.1 *	28.8
1943	48.8	37.7 *	34.2
1944	51.0	37.9 *	35.2

* includes women working part-time (2 part-time = 1 full-time)

General Consumption in Germany *per capita*, 1939–1944 (1938 = 100)	
1939	95.0
1940	88.4
1941	81.9
1942	75.3
1943	75.3
1944	70.0

Civilian Labour Force in Germany 1939–1944 (millions)					
Date	Total	German Men	German Women	Foreign Labour	Total German Workforce
1939, May 31	40.8	24.5	14.6	0.3	39.1
1940, May 31	41.6	20.4	14.4	1.2	34.8
1941, May 31	43.4	18.9	14.2	3.0	33.1
1942, May 31	44.1	16.9	14.4	4.2	31.3
1943, May 31	46.1	15.5	14.8	6.2	30.3
1944, May 31	45.2	14.2	14.8	7.1	28.9
1944, Sept 30	45.0	13.5	14.8	7.5	28.4

The Collapse of the Economy 1944–45: Coal and Transport Statistics				
Date		Coal output	Transport of coal ('000 tonnes) A = by water	B = by rail
1944	Jan	10,554	1,682	6,180
	Feb	10,482	1,835	5,820
	Mar	11,049	1,443	6,180
	Apr	10,037	2,026	5,800
	May	10,705	2,345	5,910
	Jun	10,339	2,222	6,050
	Jul	10,143	2,354	5,860
	Aug	10,417	2,214	5,650
	Sept	9,381	1,776	4,360
	Oct	7,169	724	2,520
	Nov	5,203	454	2,320
	Dec	5,370	505	2,530
1945	Jan	5,470	441	2,821
	Feb	4,678	–	2,081
	Mar	1,876	–	706

Percentage of German Industrial Labour Force Working on Orders for the Armed Forces 1939–1943

Sectors	1939	1940	1941	1942	1943
All industry	21.9	50.2	54.5	56.1	61.0
Raw materials	21.0	58.2	63.2	59.9	67.9
Metal manufacture	28.6	62.3	68.8	70.4	72.1
Construction	30.2	57.5	52.2	45.2	46.7
Consumer goods	12.2	26.2	27.8	31.7	37.0
Index (1939 = 100)	100	229	248	256	278

German State Debt Sept 1939 to May 1945 (million RM)

Date	Long and Medium-term	Treasury Bills	Tax Certificates	Other Debts	External Debt	Total Debt
01.09.39	23,910	8,340	3,230	750	1,200	37,430
31.03.40	28,667	17,731	4,110	320	1,232	52,060
31.03.41	46,566	36,123	3,654	2,104	1,212	89,649
31.03.42	69,631	61,124	3,629	6,446	1,197	142,027
31.03.43	90,898	94,736	1,244	9,557	1,185	197,620
31.03.44	118,073	144,506	1,045	10,431	1,168	275,223
31.03.45	140,000	220,000	950	11,400	1,150	373,500
08.05.45	141,000	225,500	900	11,500	1,100	388,000

Bomb Tonnages dropped on Specific Targets in Germany 1943–1945

Target	US air forces	RAF	Total
Aircraft industry	51,017	6,024	57,041
Air bases	46,979	4,353	51,332
Submarine yards	17,108	16,721	33,829
Ball bearings	6,513	13,522	20,035
Oil	130,979	93,902	224,881
Chemicals	14,615	18,212	32,827
Rubber	1,032	771	1,803
Tanks	16,922	68	16,990
Motor industry	7,378	2	7,380
Land transport	262,308	56,769	319,077
Water transport	5,100	2,233	7,333

The Effects of Bomb Attacks on Germany's Major Cities

City	Number of Major Attacks	Percentage of Built-up Area Destroyed	City	Number of Major Attacks	Percentage of Built-up Area Destroyed
Berlin	24	33	Hamburg	17	75
Bremen	12	60	Hannover	16	60
Cologne	22	61	Kassel	6	69
Dortmund	9	54	Mannheim	13	64
Dresden	1	59	Munich	9	42
Duisburg	18	48	Nuremberg	11	51
Düsseldorf	10	64	Stuttgart	18	46
Essen	28	50	Wuppertal	1	94
Frankfurt a M	11	52	Würzburg	1	89

Total tonnage dropped on city targets = 482,437

VII: The Aftermath

Refugees in the West German Provinces, 1 July 1950		
Province	Number of Refugees	as% of population
Schleswig–Holstein	915,957	34.3
Hamburg	102,714	6.4
Lower Saxony	1,842,188	26.6
North Rhine–Westphalia	1,261,391	9.5
Bremen	41,250	7.4
Hesse	680,022	15.6
Württemberg–Baden	729,101	18.6
Bavaria	1,935,504	20.9
Rhineland–Palatinate	106,093	3.6
Baden	93,098	7.0
Württemberg–Hohenzollern	109,707	8.9
Total	7,817,025	16.3

Estimated Jewish Survivors in Europe		
Country	1939 Jewish Population	Survivors, 1946
Austria	65,000	18,000
Baltic States	250,000	20,000
Belgium	90,000	25,000
Bulgaria	50,000	45,000
Czechoslovakia	315,000	60,000
Denmark	8,500	5,500
France	320,000	160,000
Germany	221,000	22,000
Greece	75,000	10,000
Hungary	400,000	220,000
Italy	50,000	30,000
Netherlands	140,000	36,000
Norway	2,000	750
Poland	3,351,000	80,000
Romania	850,000	420,000
Yugoslavia	75,000	11,000
USSR	c.5,000,000	c.4,000,000

Denazification Proceedings in the US Zone of Germany	
Total registered	13,416,101
Chargeable cases	3,669,239
Non-chargeable cases	9,746,862
Cases amnestied by Public Prosecutor	2,456,731
Cases quashed	252,875
Cases sent to trial	619,287
Persons fined	572,993
Sentenced to Special Labour	30,781
Persons barred from office	23,616
Persons restricted in employment	125,510
Persons subject to property confiscation	27,587

References

Bacque, J., *Other Losses. An Investigation into the Mass Deaths of German Prisoners of War after World War II* (London,1989)

Bagel–Bohlen, A., *Hitlers industrielle Kriegsvorbereitung im Dritten Reich 1936–1939* (Koblenz, 1975)

Barkai, A., *Nazi Economics: Ideology, Theory and Policy* (Oxford, 1990)

Baynes, N., *The Speeches of Adolf Hitler* (2 vols., Oxford, 1942)

Beck, E., *Under The Bombs: The German Home Front 1942–1945* (Lexington, Ky., 1986)

Benz, W., (ed), *Rechtsextremismus in der Bundesrepublik* (Frankfurt–am–Main, 1989)

Berghahn, V., *Modern Germany: Society, Economy and Politics in the Twentieth Century* (Cambridge, 1982)

Bloch, M., *Ribbentrop* (London, 1992)

Boelcke, W.A., *Die Kosten von Hitlers Krieg* (Paderborn, 1985)

Brady, R.A., *The Spirit and Structure of German Fascism* (London, 1937)

Brandt, K., *The Management of Agriculture and Food in the German-occupied and other areas of Fortress Europe* (Stanford, Ca., 1953)

Broszat, M., *The Hitler State* (London, 1981)

Browning, C., *The Path to Genocide: Essays on Launching the Final Solution* (Cambridge, 1992)

Bry, G., *Wages in Germany 1871–1945* (Princeton, NJ., 1960)

Burden, H.T., *The Nuremberg Party Rallies 1923–1939* (London 1967)

Burleigh, M., *Death and Deliverance: 'Euthanasia' in Germany 1900–1945* (Cambridge, 1994)

Burleigh, M., Wippermann, W., *The Racial State: Germany 1933–1945* (Cambridge, 1991)

Butler, R., *The Gestapo* (London, 1992)

Carr, W., *Germany 1815–1990* (London, 1992)

Carroll, B.A., *Design for Total War: Arms and Economics in the Third Reich* (The Hague, 1968)

Cooper, M., *The German Army 1933–1945* (London 1978)

Cormi. G., *Hitler and the Peasants* (Oxford, 1990)

Dallin, A., *German Rule in Russia* (2nd ed., London, 1981)

Dear, I.C.B., Foot, M.R.D., *The Oxford Companion to the Second World War* (Oxford, 1995)

Deist, W., et al., *Germany and the Second World War, Vol. 1* (Oxford, 1990)

Diefendorf, J., *In the Wake of War: the Reconstruction of German Cities after World War II* (London, 1993)

Domarus, M., *Hitler. Reden und Proklomationen 1933–1945* (2 vols., Munich, 1965)

Dülffer, J., *Weimar, Hitler und die Marine: Reichspolitik und Flottenbau* (Düsseldorf, 1973)

Eatwell, R., *Fascism: a History* (London, 1995)

Eichholtz, D., *Geschichte der deutschen Kriegswirtschaft. Vol I, 1939–1941* (Berlin, 1969); *Vol II, 1941–1943* (Berlin, 1985)

Engelmann, B., *In Hitler's Germany: Everyday Life in the Third Reich* (London, 1988)

Falter, J., *Hitlers Wähler* (Munich, 1991)

Farquharson, J., *The Plough and the Swastika: the NSDAP and Agriculture in Germany 1928–1945* (London, 1976)

Feldman, G., *The Great Disorder: Politics, Economics and Society in the German Inflation 1914–1924* (Oxford, 1993)

Fischer, C., *The Rise of the Nazis* (Manchester, 1995)

Fowkes, B., *Communism in Germany under the Weimar Republic* (London, 1984)

Freeman, M., *Atlas of Nazi Germany* (London, 1987)

Frei, N., *National Socialist Rule in Germany: the Führer State 1933–1945* (Oxford 1993)

Gellately, R., *The Gestapo and German Society: Enforcing Racial Policy 1933–1945* (Oxford, 1990)

Georg, E., *Die wirtschaftliche Unternehmungen der SS* (Stuttgart, 1963)

Gilbert, F., (ed.), *Hitler Directs his War* (Oxford, 1950)

Gillingham, J., *Industry and Politics in the Third Reich* (London, 1985)

Grau, G., *Hidden Holocaust? Gay and Lesbian Persecution in Germany 1933–1945* (London, 1995)

Grill, J., *The Nazi Party in Baden 1920–1945* (Chapel Hill, NC, 1983)

Grunberger, R., *A Social History of the Third Reich* (London 1971)

Guillebaud, C.W., *The Economic Recovery of Germany 1933–1938* (London, 1939)

Hart–Davis, D., *Hitler's Games: the 1936 Olympics* (London 1986)

Herbert, U., *Fremdarbeiter: Politik und Praxis des 'Ausländer–Einsatzes' in der Kriegswirtschaft des Dritten Reiches* (Bonn, 1985)

Herzstein, R., *When Nazi Dreams Come True: A Look at the Nazi Mentality 1939–1945* (London, 1982)

Hilberg, R., *Perpetrators, Victims, Bystanders: The Jewish Catastrophe 1933–1945* (London, 1993)

Hilgemann, W., *Atlas zur deutschen Zeitgeschichte 1918–1968* (Munich, 1984)

Hirsch, K., *Rechts: aktuelles Handbuch zur rechtsextremen Szene* (Berlin, 1990)

Holtfrerich, C., *The German Inflation 1914–1923* (Berlin, 1986)

Homze, E., *Foreign Labor in Nazi Germany* (Princeton, N.J., 1967)

Hubert, P., *Uniformierter Reichstag: die Geschichte der Pseudo–Volksvertretung 1933–1945* (Düsseldorf, 1992)

James, H., *The German Slump: Politics and Economics 1924–1936* (Oxford, 1986)

Kaiser, D., *Economic Diplomacy and the Origins of the Second World War* (Princeton, N.J., 1980)

Kater, M., *The Nazi Party. A Social Profile of Members and Leaders 1919–1945* (Oxford, 1983)

Keegan, J., (ed.) *The Times Atlas of the Second World War* (London, 1989)

Kershaw, I., *The Nazi Dictatorship* (London, 1985)

Klein, B.H., *Germany's Economic Preparations for War* (Cambridge, Mass., 1959)

Kramer, A., *The West German Economy 1945–1955* (Oxford, 1991)

Madej, W.V., *The German War Economy: the Motorization Myth* (Atlanta, Pa., 1984)

Maier, K., et al., *Germany and the Second World War: Vol. II* (Oxford, 1991)

Manvell, R., Fraenkel, H., *The July Plot* (London, 1964)

Maser, W., *Hitler's Letters and Notes* (London, 1974)

McIsaac, D., (ed.), *The United States Strategic Bombing Survey* (8 vols., New York, 1980)

Mierzejewski, A.C., *The Collapse of the German War Economy 1944–1945* (Chapel Hill, N.C., 1987)

Milward, A.S., *The German Economy at War* (London, 1965)

Milward, A.S., *The New Order and the French Economy* (Oxford, 1970)

Nicosia, F., *The Third Reich and the Palestine Question* (Austin, Tex., 1985)

Noakes, J., (ed.), *Government, Party and People in Nazi Germany* (Exeter, 1980)

Orlow, D., *The History of the Nazi Party* (2 vols., Newton Abbot, 1969)

Overy, R.J., *The Nazi Economic Recovery 1932–1938* (2nd ed., Cambridge, 1996)

Overy, R.J., *War and Economy in the Third Reich* (Oxford, 1994)

Overy, R.J., *Why the Allies Won* (London, 1995)

Owings, A., *Frauen: German Women Recall the Third Reich* (London, 1993)

Petzina, D., *Autarkiepolitik im Dritten Reich* (Stuttgart, 1968)

Peukert, D., *Inside Nazi Germany: Conformity, Opposition and Racism in Everyday Life* (London, 1987)

Pounds, N., *The Economic Pattern of Modern Germany* (London, 1963)

Reuth, R.G., *Goebbels* (London, 1993)

Rich, N., *Hitler's War Aims* (2 vols., London 1973–4)

Royal Institute of International Affairs, *Hitler's Europe* (Oxford, 1954)

Schliephake, H., *The Birth of the Luftwaffe* (London, 1971)

Schneider, C., *Stadtgründung im Dritten Reich: Wolfsburg und Salzgitter* (Munich, 1979)

Schoenbaum, D., *Hitler's Social Revolution. Class and Status in Nazi Germany 1933–1939* (New York, 1966)

Siegel, T., 'Wage Policy in Nazi Germany' *Politics and Society 14* (1985)

Smelser, R., *Robert Ley: Hitler's Labor Front Leader* (London, 1988)

Snyder, L., *Encyclopedia of the Third Reich* (London, 1976)

Speer, A., *Inside the Third Reich* (London, 1970)

Stachura, P., *Nazi Youth in the Weimar Republic* (Oxford, 1975)

Steinert, M., *Hitler's War and the Germans* (Athens, Ohio, 1977

Stephenson, J., *Women in Nazi Society* (London, 1975)

Stüber, G., *Der Kampf gegen den Hunger 1945–1950* (Neumünster, 1984)

Teichert, E., *Autarkie und Grossraumwirtschaft in Deutschland 1930–1939* (Munich, 1984)

Thies, J., *Architekt der Weltherrschaft: Die Endziele Hitlers* (Düsseldorf, 1976)

Trevor–Roper, H., (ed.), *The Goebbels Diaries: the Last Days* (London, 1978)

Trevor–Roper, H., (ed.), *Hitler's War Directives 1939–1945* (London, 1964)

Turner, I. (ed.), *Reconstruction in Postwar Germany* (Oxford, 1989)

Wagner, J.F., *Brothers Beyond the Sea: National Socialism in Canada* (Waterloo, Ont., 1981)

Webster, C., Frankland, N., *The Strategic Air Offensive Against Germany 1939–1945* (4 vols., London, 1961)

Welch, D., (ed.), *Nazi Propaganda* (London, 1983)

Wendt, B–J., *Grossdeutschland: Autarkiepolitik und Kriegsvorbereitung des Hitler–Regimes* (Munich, 1987)

Werner, W.F., *Bleib übrig! Deutscher Arbeiter in der ns Kriegswirtschaft* (Düsseldorf, 1983)

Zeman, Z., *Nazi Propaganda* (2nd ed., Oxford, 1973)

Zumpe, L., *Wirtschaft und Staat in Deutschland, 1933–1945* (Berlin, 1980)

Index

Acknowledgements

Picture Credits

Front Cover

(Clockwise from top left)

Blitzkrieg: German Invasion of the Low Countries, May 1940, German troops advance into battle

Hitler at Nazi Rally in Dortmund, 1933

WWII German troops during the invasion of Poland in 1939

Nazi Germany: Brownshirts parade with flags on Party Day at Nuremburg, 1933

Troops of the 1st Division of the 8th German Army Corps march through Vienna after the German occupation of Austria in March, 1938

"Bombing Up" loading junkers Ju 87 Stuka dive-bombers during the Battle of Britain, 1940

20th Century Germany: Adolf Hitler and Heinrich Himmler, head of SS in 1938

Hitler's Mein Kampf partwork published by Hutchinson, London, 1940

(All the above pictures were used with courtesy from the Collection of Peter Newark's Historical Pictures)

Internal

Author's Collection: 97b

Courtesy of the Institute of Contemporary History and Wiener Library: 90, 92

Peter Newark's Historical Pictures: 8l, 8b, 9tl, 9bl, 10, 11, 12, 15, 16, 18, 21, 22, 23tl, 23bl, 24tl, 24bl, 25, 27, 28, 31, 33, 34, 35, 36, 38, 41, 42, 43, 44, 46, 47, 48tl, 48b, 49, 50, 52, 55, 56, 59, 60, 61, 62, 63t, 63b, 65, 67, 68, 71, 72l, 73bl, 75, 77, 78, 79t, 79b, 80, 81, 83, 84, 87r, 87tl, 94, 95, 96, 97t, 99, 101, 103, 105, 106, 107, 108t, 108b, 109, 112

FOR SWANSTON PUBLISHING LIMITED

Concept:
Malcolm Swanston

Editorial and Map Research:
Andrew and Ailsa C. Heritage

Editorial Assistance:
Stephen Haddelsey

Cartography:
Andrea Fairbrass
Elsa Gibert
Kevin Panton

Index:
Jean Cox
Barry Haslam

Typesetting:
Andrea Fairbrass
Charlotte Taylor

Picture Research:
Charlotte Taylor

Production:
Advanced Illustration Ltd, Congleton, UK.
Barry Haslam

Separations:
Advanced Illustration Ltd, Congleton, UK.